About Island Press

Since 1984, the nonprofit Island Press has been stimulating, shaping, and communicating the ideas that are essential for solving environmental problems worldwide. With more than 800 titles in print and some 40 new releases each year, we are the nation's leading publisher on environmental issues. We identify innovative thinkers and emerging trends in the environmental field. We work with world-renowned experts and authors to develop cross-disciplinary solutions to environmental challenges.

Island Press designs and implements coordinated book publication campaigns in order to communicate our critical messages in print, in person, and online using the latest technologies, programs, and the media. Our goal: to reach targeted audiences—scientists, policymakers, environmental advocates, the media, and concerned citizens—who can and will take action to protect the plants and animals that enrich our world, the ecosystems we need to survive, the water we drink, and the air we breathe.

Island Press gratefully acknowledges the support of its work by the Agua Fund, Inc., The Margaret A. Cargill Foundation, The Nathan Cummings Foundation, Betsy and Jesse Fink Foundation, The William and Flora Hewlett Foundation, The Kresge Foundation, The Forrest and Frances Lattner Foundation, The Andrew W. Mellon Foundation, The Curtis and Edith Munson Foundation, The Overbrook Foundation, The David and Lucile Packard Foundation, The Summit Foundation, The Summit Fund of Washington, Trust for Architectural Easements, The Winslow Foundation, and other generous donors.

The opinions expressed in this book are those of the author(s) and do not necessarily reflect the views of our donors.

BEYOND NATURALNESS

Beyond Naturalness

RETHINKING PARK AND WILDERNESS STEWARDSHIP IN AN ERA OF RAPID CHANGE

EDITED BY

David N. Cole and Laurie Yung

Washington | Covelo | London

No copyright claim is made in the work of David N. Cole, Paul Deprey, David M. Graber,
Peter Landres, Constance I. Millar, David J. Parsons, Nathan L. Stephenson,
Kathy A. Tonnessen, and Leigh A. Welling, employees of the federal government.

Library of Congress Cataloging-in-Publication Data

Beyond naturalness : rethinking park and wilderness stewardship in an era of rapid change /
edited by David N. Cole and Laurie Yung.
p. cm.
Includes bibliographical references and index.
ISBN-13: 978-1-59726-508-9 (cloth : alk. paper)
ISBN-10: 1-59726-508-X (cloth : alk. paper)
ISBN-13: 978-1-59726-509-6 (pbk. : alk. paper)
ISBN-10: 1-59726-509-8 (pbk. : alk. paper)
1. Protected areas–United States–Management. 2. Ecosystem management–United States. 3.
Nature conservation–United States. I. Cole, David N. II. Yung, Laurie, 1969-
S930.B49 2010
333.78–dc22
2009038811

Printed on recycled, acid-free paper ♻

Manufactured in the United States of America

10 9 8 7 6 5 4 3 2

CONTENTS

Parks, wilderness, and other protected areas have been established to protect biodiversity and an array of natural and cultural values. However, it is increasingly clear that designation of protected areas—drawing lines around particular places and letting them be—will not preserve everything we value in these places. Threats to protected area ecosystems are mounting, forcing managers to implement or at least consider active ecosystem management.

Our interest in producing this book stemmed from two ultimately converging experiences. We are deeply concerned about the increasingly pressing threat that a changing climate poses for all life on Earth, human and nonhuman, and the special role protected areas play in preserving biodiversity as we move into and, we hope, through these unprecedented times. In many senses, climate change and other novel environmental stressors require us to think differently about the human relationship with nature and therefore about how we manage and conserve protected areas. With this in mind, we found ourselves questioning some of the sacred tenets of park and wilderness management and looking for new ways to secure the future of these extraordinary places.

Our interest in the book was also spurred by recognition that the nature of public land management in the United States has changed substantially. Compared with those of the past, today's park and wilderness managers are held to much higher standards regarding how they make management decisions. Doing what they think is right, without justifying what they are doing and why, is no longer acceptable. Instead, before acting, planners and managers must describe objectives, desired future conditions, and the outcomes of management actions in as specific terms as possible. The public is able to participate in development of these specifications and, because objectives are transparent, able to monitor the success of management and hold agencies accountable.

Developing specific statements regarding management outcomes is challenging. One of us (David Cole) has worked with managers to develop more specific operational objectives, largely in the context of recreation management; however, it has proven difficult to translate what worked for managing recreation to managing ecosystems, which are much more complex and dynamic. In 1996, David attended a workshop on the concept of historical range of variability (HRV), which was being promoted in part as a means of articulating desired future conditions and management targets. Little attention at that workshop was devoted to how directional global change might make HRV a misleading guide for defining desired future conditions. Since then, however, increased attention to climate change has made such concerns more obvious and widely accepted.

As we thought about the challenges of conserving parks and wilderness in the face of global change and the need to articulate specific objectives related to whether, when, and how we should intervene in ecosystem processes, several questions arose. Is the concept of naturalness, the most fundamental management objective for parks and wilderness, part of the problem? Do we need to look beyond naturalness for concepts that might provide a better foundation for achieving conservation objectives? In seeking answers to these questions, we gathered together a group of scientists and scholars we deeply respect, an interdisciplinary team drawn from such fields as ecology, wildlife biology, paleoecology, geography, social science, and philosophy. Project participants came from academia, federal agencies, and the conservation community and from the United States, Canada, and Australia. They brought experience and expertise in working with fire, invasives, climate change, ethics, policy, restoration, wilderness, and national parks and, perhaps most important, the ability to synthesize, integrate, and innovate across these fields. In 2007, we convened the first of two workshops in western Montana to initiate a dialogue that has shaped and reshaped our thinking about these pressing issues. Our goal in producing this book is to broaden that dialogue, to engage managers, scientists, conservationists, and the public in a discussion about how to conserve these special places into the future.

We are deeply grateful to everyone who made this book possible. Our editors at Island Press, Barbara Dean, Barbara Youngblood, and Erin Johnson provided sage advice, unending patience, and kind encouragement. Twenty-six anonymous reviewers supplied critical feedback on each of the chapters. The Aldo Leopold Wilderness Research Institute, Rocky Mountain Research Station, USDA Forest Service; the University of Montana Wilderness Institute; and the University of Montana College of Forestry and Conservation provided funding and staff support for this project. We are forever indebted to all the project participants (the contributors to this volume) who dedicated their time and intellect to this project and this book. Their wisdom, insight, commitment, hard work, and willingness to passionately engage in dialogue at the conference room table, on the porch of the cabin, on the backcountry trail, and in the pages of this text truly made this book possible.

We want to express our appreciation for the many special places that have inspired us to care—places too numerous to list but including Walling Reef, Grand Canyon, Bowman Lake, the Sierra Nevada, Neon Canyon, the Eagle Cap, and Bighorn Crags. Finally, we want to acknowledge the countless dedicated people who manage and work to protect these places. In the pages that follow, we challenge park and wilderness managers to develop new perspectives and try new approaches. We ask for change respectfully, mindful that implementation is challenging and that the burden of the challenge lies with protected area managers rather than us.

David N. Cole and Laurie Yung

Chapter 1

Park and Wilderness Stewardship: The Dilemma of Management Intervention

DAVID N. COLE AND LAURIE YUNG

> We can't solve problems by using the same kind of thinking we used when we created them.
>
> —*Albert Einstein*

National parks and wilderness areas are an important part of America's natural and cultural heritage. On the surface at least, people appear to share a common vision of park purposes and wilderness values, believing that parks and wilderness are places set aside and protected from development to preserve their beauty, their natural features, and the opportunity for future generations to learn from them, love them, and experience them as we have. In addition to being powerful symbols that spark the imagination, parks and wilderness play a critical role in environmental conservation, especially as global threats to biodiversity mount.

From global climate change and invasive species to pollution and land fragmentation, anthropogenic stressors threaten park and wilderness values and raise serious questions about what it means to preserve our natural heritage. We cannot preserve parks and wilderness by drawing a line around them and leaving them alone. Protecting an area's beauty, heritage, and biodiversity entails thoughtful stewardship and, at times, active intervention. But active intervention presents a new set of challenges. Do park and

1

wilderness managers have the policy guidance they need to be effective land managers in this changing context?

In this book we explore the goals that guided the conservation of large protected areas in the twentieth century, most of them related to the concept of naturalness. These goals were appropriate a century ago, when the struggle was one of protecting land from development and exploitation, and they retain iconic meaning and value today. But over the past century the world has changed and the pace of change has accelerated. The most certain characteristic of the future is uncertainty. The stewardship issues of the twenty-first century will be more nuanced, with solutions that are less clear cut, less black and white. Consequently, it is time to think beyond naturalness, to articulate park purposes in terms that are both more specific and more diverse than naturalness and to adopt a wider array of management approaches to achieve these purposes.

The concept of naturalness is more central to the stewardship of national parks and wilderness in the United States than to stewardship of other types of protected areas or parks and wilderness in other countries. Therefore, this book is focused on parks and wilderness in the United States. However, many of the issues raised in this book have broad international application. To set the stage for the chapters that follow, this introduction first explores some of the emerging dilemmas confronting park and wilderness managers. We then describe the central argument of the book and provide a guide to the chapters that follow.

Stewardship Dilemmas

Stretching from the banks of the Rio Grande up desert canyons to the forested slopes of the Jemez Mountains, not far from Santa Fe, New Mexico, is Bandelier National Monument. One of the oldest units of the National Park System, Bandelier was established in 1916. In 1976, most of the park (more than 23,000 acres) was designated as wilderness. Ancestral Pueblo people inhabited these lands long ago, in substantial numbers from at least the twelfth through the sixteenth centuries. Remnants of their life here — from cliff dwellings to painted caves and potsherds — draw many visitors, as do the diverse desert, woodland, and forest landscapes. The monument is a place of peace and solace, and it provides connections to the past and to wild and natural landscapes.

All is not well here, however. Despite a long history of human habitation and the fact that Bandelier has not been a pristine landscape for more

than a thousand years, more recent Euro-American use of the park landscape
has triggered unprecedented change in most of the park's ecosystems (Allen 2004). Conditions, particularly in the wilderness' piñon–juniper woodlands, have rapidly deteriorated. Studies of current and historical ecology
tell a sad story (Sydoriak et al. 2000). Herbaceous ground cover and surface
organic litter have largely disappeared, exposing soils to the erosive effects
of rainfall, which can come in torrents during the summer monsoons. The
park's woodland soils, which developed over tens of thousands of years, are
being washed away rapidly. Most soil will be gone within about a century,
with much associated loss of the artifacts and cultural heritage that were the
primary reasons the monument was originally established.

As detailed by Sydoriak et al. (2000) and Allen (2004), these changes
are the result of increases in the density and canopy closure of piñon and
juniper trees—a response to the cumulative and synergistic effects of human activities, along with natural events. Intense livestock grazing, which
occurred from the 1880s until 1932, resulted in loss of much of the herbaceous ground cover. Piñon–juniper density had long been limited by frequent, wide-ranging surface fires. The absence of an herbaceous ground
cover to carry fire—along with federal programs to suppress fire—meant
that fires no longer checked tree establishment. Consequently, twentieth-
century rates of tree growth were unprecedented. More trees means even
less herbaceous growth because trees effectively compete for water and nutrients, creating a positive feedback loop that favors further tree invasion
and decreased herbaceous ground cover (Sydoriak et al. 2000). To add in-
sult to injury, feral burros from about 1930 to 1980 and severe drought in
the 1950s further contributed to declines in ground cover.

Because grazing and fire suppression caused the problem at Bandelier,
one solution might be to eliminate grazing and restore fire. Indeed, this is
part of the solution, but it is not enough. Unfortunately, as in many other
protected areas, ecological thresholds have been crossed that are not so easy
to renavigate (Allen 2007). The ecosystem has been altered so fundamentally that eliminating the original cause of the problem, or simply leaving
the system alone to restore itself, no longer resolves the situation. According to ecologists working at Bandelier, the solution is not less but more
human intervention. Experiments have demonstrated that if small trees are
cut down over large swaths of land, a more continuous ground cover of
native herbs and grasses can be restored, which in turn will reduce erosion
and the loss of cultural artifacts. As Sydoriak et al. (2000: 212) observe,
"This treatment directly reduces tree competition with herbaceous plants
for scarce water and nutrients, and the application of slash residues across

the barren interspaces greatly reduces surface water runoff and ameliorates the harsh microclimate at the surface, immediately improving water availability for herbaceous plants." After years of research, experimentation, and detailed assessment of trade-offs, Bandelier is implementing "ecological restoration" treatments across large portions of the park's designated wilderness (National Park Service 2007).

Forest thinning may be a solution to environmental degradation, but at what cost? Isn't wilderness supposed to be a place where interventions such as cutting trees are not allowed—where nature decides what's best? After all, current problems are the direct result of earlier generations doing what they thought was best. Is further intervention just a perpetuation of this arrogance? Such thinking led Howard Zahniser (1963), principal architect of the Wilderness Act, to famously declare that, when it comes to wilderness, we should be guardians, not gardeners. Or is it our responsibility to correct our past mistakes—to facilitate the healing process—and help Bandelier's ecosystems transition to a more functional state? Should we listen not to Zahniser but to acclaimed biologist Dan Janzen (1998) and embrace our role as gardeners, as the only responsible way to sustain wildlands into the future?

Such dilemmas—and the questions they raise—are all too common in parks and wilderness these days. On the other side of the continent, on national forest land in the state of Virginia, is the Saint Mary's Wilderness. It was once one of the most pristine watersheds in the region (Figure 1.1), but long-term monitoring revealed a sobering trend of environmental degradation. As reported in an environmental assessment, by the late 1990s the pH of the Saint Mary's River had declined from historic levels of about 6.8 to levels of 4.0–5.6—a more than 100-fold increase in acidity. The diversity of aquatic macroinvertebrates—the foundation of the aquatic food chain—had declined 38 percent. The number of fish species had declined from twelve to just three, and two of those remaining species—brook trout (*Salvelinus fontinalis*) and black-nosed dace (*Rhinichthys atratulus*)—had not reproduced successfully for several years. The culprit was air pollution, in the form of sulfur and nitrogen compounds, being produced both locally and in the Ohio Valley.

In 1999, wilderness managers responded. A helicopter dumped 140 tons of limestone sand at six locations in the wilderness, adjacent to the river and its tributaries. This buffered 12 miles of stream in this small (7,000 acre) wilderness. Within 200 days, stream pH was back to 6.5 and macroinvertebrate diversity and fish density had increased dramatically. Was this intervention a success? By many measures, one would have to say yes.

FIGURE I.I. The Saint Mary's River in the Saint Mary's Wilderness has been adversely affected by air pollution. Managers have responded by periodically dumping lime close to the river, raising its pH. (Photo by Steven Brown)

But as in Bandelier, one must ask whether the ends justify the means. Does more human intervention make things better or worse? Can two wrongs make a right? Moreover, in this case the proposed response did not attack the root cause of the problem: the production of atmospheric pollutants. It merely treated symptoms in the hope that a more permanent solution to the problem could be found. Success was temporary; 6 years later the treatments had to be repeated.

In the Big Gum Swamp Wilderness in northern Florida, fires historically burned through pine and wiregrass communities every 3 to 5 years (Christensen 1978). Fires seldom ignited within the boundaries of the small (13,660 acre) wilderness. However, fire frequency was high because, in the past, fires swept unimpeded across large portions of northern Florida's undeveloped, unfragmented landscape. Fires ignited by distant lightning strikes often burned into and across the wilderness. In recent decades, development around the wilderness has made it an isolated island, cut off from historic disturbance processes. Fire occurs infrequently.

Consequently, vegetation composition and structure have changed, fuels have built up, and the potential for catastrophic fire has increased. Biodiversity is being lost. Wilderness managers have responded by mimicking, in part, the natural fire regime, intentionally setting fires inside the wilderness every few years.

Some of these interventions, undertaken to respond to undesirable human impact, may be needed at extremely large spatial scales. For example, whitebark pine (*Pinus albicaulis*) is the dominant timberline tree species of many of the parks and wilderness areas of the West, from the northern Rocky Mountains to the Sierra Nevada and Cascade Mountains. Throughout much of its range, whitebark pine is being decimated by white pine blister rust (*Cronartium ribicola*), a nonnative fungal pathogen inadvertently introduced in 1910. Mountain pine beetles (*Dendroctonus ponderosae*) are also killing whitebarks. In combination, blister rust and mountain pine beetles have killed more than 50 percent of whitebark pine across much of the northern part of its range in the United States, including parks such as Glacier National Park (Kendall and Keane 2001). Severe whitebark pine mortality could seriously affect grizzly bear (*Ursus arctos horribilis*) populations, which depend on pine seeds for a significant portion of their diet (Mattson et al. 1991). Whitebark mortality is exacerbated by decades of fire suppression, which has led to successional replacement of the pine by more shade-tolerant tree species. And yet the reintroduction of fire is also problematic for whitebark because decimated stands do not produce enough seed to naturally regenerate after disturbance (Tomback et al. 2001). Potential solutions include breeding rust-resistant trees, planting seedlings, cutting trees to create a more diverse and resistant age class structure, and burning—actions that may be taken across large swaths of park and wilderness lands.

There are an endless number and variety of challenging dilemmas like those described here. In the mountainous parks and wilderness of the eastern United States, hemlock (*Tsuga*) species are being decimated by the hemlock wooly adelgid (*Adelges tsugae*), an introduced aphid-like insect. Here the intervention involves insecticide injections and the introduction of nonnative beetles that prey on the adelgid. In the canyon country of the southwestern deserts—places such as Grand Canyon National Park— riverbanks are now dominated by the introduced shrub tamarisk (*Tamarix* sp.). The obvious solution is to eliminate the tamarisk and restore the natural ecosystem. But other values must be considered. Tamarisk thickets have become favored and important habitat for sensitive bird species such as Bell's vireo (*Vireo bellii*) and the endangered southwestern willow flycatcher (*Empidonax traillii extimus*)—species that have lost most of their original habitat to rampant development throughout the southwestern states. Which is more important: reducing the risk of species extinction or restoring riverbanks to a previous state?

Wilderness and national parks around the world are experiencing similar stewardship dilemmas. In Kruger National Park, South Africa, burgeoning populations of elephants confined within park boundaries are converting woodland into grassland, forcing park managers to propose culling. We have been slow to recognize such conundrums and even slower to act on them. Action is difficult because it raises ethical concerns and requires value-based, often highly political decisions. Should park and wilderness managers intervene in natural biological and physical processes? If so, under what circumstances and how? What specific actions should managers take? Unless predictions about the likely consequences of climate change are grossly overstated, climate change is about to hone the horns of this dilemma to the point where it can no longer be ignored. Some scientists (Cole et al. 2005) predict that, given a changing climate, Joshua trees (*Yucca brevifolia*) may no longer be capable of surviving unaided in Joshua Tree National Park. What does this suggest about park purposes and appropriate responses? Should the park follow the Joshua trees? Will the trees be able to migrate successfully? Does the park have an obligation to help secure the future persistence of Joshua trees on lands outside park boundaries? What are the conservation goals of a park that has lost its icon and signature botanical element?

Guidance for Management Interventions

As the preceding examples illustrate, the key challenge to park and wilderness stewardship is to decide where, when, and how to intervene in physical and biological processes to conserve what we value in these places. Some of what we call intervention and active management involves ecological restoration, the process of assisting the recovery of ecosystems that have been damaged, degraded, or destroyed (Society for Ecological Restoration International 2006). But we prefer the more generic term *intervention* because it includes any prescribed course of action that intentionally alters ecosystem trajectories and avoids the connotation of a return to past conditions. In many cases, *redirection* might be a better term than *restoration*. Interventions range from lighting fires to culling ungulate populations, from thinning forests to assisting migration of individuals or species better adapted to changing conditions. Some are one-time actions, such as introducing a species and stepping back to see whether it can thrive in a new site. Others are ongoing, such as the liming of the Saint Mary's River. Some interventions are small in scale (e.g., actively maintaining a 10-acre forest of sequoia [*Sequoiadendron giganteum*] or Joshua tree woodland at a location no longer ideal for the species), whereas others might be large in scale (e.g., burning tens of thousands of acres each year).

This book traces how park goals and purposes have diversified and become both more complex and contested. It was originally assumed that the varied purposes for which parks and wilderness areas were established were internally consistent and could all be subsumed under the central purpose of preserving natural conditions. Managing for naturalness was thought to be an effective way to preserve all the objects and values to be conserved in parks and wilderness. But, as this book argues, it is becoming increasingly clear that no single management approach can protect and preserve the full range and diversity of park and wilderness purposes and values. Trade-offs are necessary.

We argue that the goals that guided the conservation and restoration of large protected areas in the twentieth century—most notably the concept of naturalness—do not provide sufficient guidance for park and wilderness stewardship. Regarding the decision of whether to intervene, the concept of naturalness is not helpful. *Intervention* implies exerting human control to compensate for human impact on the land. Because *naturalness* implies both a lack of human impact and a lack of human control, one of the meanings of *naturalness* will be violated whatever is done (or not done). Where interventions are pursued, decisions must be made about how to intervene, and well-supported management objectives and desired outcomes must be articulated. Objectives and outcomes must be knowable, attainable, and desirable. By most definitions, objectives based on naturalness have none of these attributes.

Given the stewardship challenges that managers currently face and changes that are already occurring, such as climate change and habitat fragmentation, we believe it is time to rethink park and wilderness goals and the assumptions on which they rest. It is increasingly clear that just leaving nature alone will not be adequate to conserve biodiversity and many of the other values we associate with protected areas. Our goal in developing this book is to engage management agencies, scientists, policymakers, and the public in careful deliberation about the future of protected areas in the United States and what we, as a society, want to conserve and protect in these places. We believe that such a dialogue will result in more explicit and transparent consideration of priorities and trade-offs and, ultimately, more innovative and effective strategies for adapting to the changing context that climate change and a host of other environmental stressors now present.

The Chapters That Follow

As described in this chapter, the aim of this book is to increase awareness about limitations of traditional protected area goals, advance alternative concepts that might be useful, and suggest some specific management strategies and innovations to try. Our focus is on the management of the ecosystems and ecological elements of parks and wilderness, although we recognize that protected areas serve other purposes, such as education, recreation, and livelihood. We do not attempt to resolve the long-standing tension between use and preservation. Regarding terminology, we use *management* and *stewardship* interchangeably, always for the purpose of conservation of what is valued in parks and wilderness. We also use the more specific terms *parks* and *wilderness* and the broader term *protected area* interchangeably. As described in this chapter, although this book focuses on U.S. parks and wilderness, the protected areas that have the clearest mandate to preserve naturalness, many of the concepts and approaches described herein are applicable to other types of protected areas in the United States and abroad.

The book consists of two introductory chapters, three parts, and a conclusion. Chapter 2 explores naturalness, the concept that is central to park and wilderness policy and management. This is followed by the first part of the book, "The Changing Context of Park and Wilderness Stewardship," which reviews the changes that compel us to rethink and better articulate the goals, purposes, and strategies of protected area conservation. Chapter 3 describes how scientific understanding of ecosystems has evolved and the implications of new ecological paradigms for protected area management. Chapter 4 explores our understanding of future environmental change and biotic response to change, with a focus on climate change and multiple ecosystem stressors, including habitat fragmentation, invasive species, and pollution. Chapter 5 examines how agencies have struggled to put naturalness policies into practice and the problematic nature of vague and ambiguous policy direction.

The second part, "Approaches to Guide Protected Area Conservation," explores a variety of approaches to the conservation of parks and wilderness. Just as there are a number of different park purposes, there are a variety of management approaches, which vary in their likely outcomes.

Collectively, park goals and purposes are likely to be optimized through a diversity of approaches. Chapter 6 describes a hands-off, nonintervention approach to stewardship that would leave nature alone, an approach well fitted to the purpose of protecting nature's autonomy. Chapter 7 describes an approach that emphasizes ecological integrity, using active management to maintain intact, sound, properly functioning ecosystems. Chapters 8 and 9 advance the concepts of historical fidelity and resilience and explore how these concepts might change the way protected area ecosystems are managed. Although there is substantial overlap among these four approaches, each differs in its central emphasis.

The third part, "Management Strategies for Implementing New Approaches," provides more specific strategies for responding to changing conditions and stewardship dilemmas. Chapter 10 explores the importance to stewardship programs of developing realistic objectives and setting priorities. Invasive species management is used as an example because there is no question that future ecosystems will include invasive species. Chapter 11 focuses specifically on management responses to climate change, increasing resistance and resilience, adaptation, and realignment. Chapter 12 describes the increased importance of planning for conservation at larger spatial scales (landscapes and systems of protected areas), and Chapter 13 explores the implications of change and uncertainty to planning frameworks and institutions. Chapter 14 advances the concept of wild design, acknowledging the intentionality inherent in interventions but balancing it with reciprocity and ecosystem knowledge. The book concludes with a chapter that synthesizes previous material and explores possible paths forward, with a focus on institutional change, policy options, and scientific research.

REFERENCES

Allen, C. D. 2004. Ecological patterns and environmental change in the Bandelier landscape. Pp. 19–68 in T. A. Kohler, ed. *Village formation on the Pajarito Plateau, New Mexico: Archaeology of Bandelier National Monument*. University of New Mexico Press, Albuquerque.

Allen, C. D. 2007. Cross-scale interactions among forest dieback, fire, and erosion in northern New Mexico landscapes. *Ecosystems* 10:797–808.

Christensen, N. L. 1978. Fire regimes in southeastern ecosystems. Pp. 112–16 in *Fire regimes and ecosystem properties: Proceedings of the conference*. General Technical Report WO-26. USDA Forest Service, Washington, DC.

Cole, K. L., K. Ironside, P. Duffy, and S. Arundel. 2005. *Transient dynamics of vegetation response to past and future climatic changes in the southwestern United States.*

Poster presented at the workshop of the U.S. Climate Change Science Program, Arlington, VA, November 14–16, 2005. Retrieved April 2, 2008 from www.climatescience.gov/workshop2005/posters/P-EC4.2_Cole.pdf.

Janzen, D. H. 1998. Gardenification of wildland nature and the human footprint. *Science* 279:1312–1313.

Kendall, K. C., and R. E. Keane. 2001. Whitebark pine decline: Infection, mortality, and population trends. Pp. 221–242 in D. F. Tomback, S. F. Arno, and R. E. Keane, eds. *Whitebark pine communities: Ecology and restoration*. Island Press, Washington, DC.

Mattson, D. J., B. M. Blanchard, and R. R. Knight. 1991. Food habits of Yellowstone grizzly bears, 1977–1987. *Canadian Journal of Zoology* 9:1619–1629.

National Park Service. 2007. *Bandelier National Monument: Final ecological restoration plan and environmental impact statement*. Retrieved March 17, 2009 from parkplanning.nps.gov/.

Society for Ecological Restoration International Science and Policy Working Group. 2006. *The SER International primer on ecological restoration*. Society for Ecological Restoration International, Tucson, AZ.

Sydoriak, C.A., C. D. Allen, and B. Jacobs. 2000. Would ecological landscape restoration make the Bandelier Wilderness more or less of a wilderness? Pp. 209–215 in D. N. Cole, S. F. McCool, W. T. Borrie, and J. O'Loughlin, comps. *Wilderness Science in a Time of Change conference*, Volume 5: *Wilderness ecosystems, threats and management*. Proceedings RMRS-P-15-VOL-5. USDA Forest Service, Rocky Mountain Research Station, Ogden, UT.

Tomback, D. F., S. F. Arno, and R. E. Keane, eds. 2001. *Whitebark pine communities: Ecology and restoration*. Island Press, Washington, DC.

Zahniser, H. 1963. Guardians not gardeners. *The Living Wilderness* 83:2.

Chapter 2

The Trouble with Naturalness: Rethinking Park and Wilderness Goals

GREGORY H. APLET AND DAVID N. COLE

> Words can assume quite different meanings as time passes, as context changes, or even as they are spoken by different people. In resource management, the interpretation of a few key phrases has caused and continues to cause untold havoc.
>
> —*Luna Leopold*

Naturalness, more commonly phrased as "natural conditions," appears as a guiding concept throughout protected area policy. The National Park Service Organic Act of 1916 declared that the fundamental purpose of the parks was "to conserve the scenery and the natural and historic objects and the wild life therein . . . unimpaired for the enjoyment of future generations." Historian Richard Sellars (1997) argues that the provision that parks remain "unimpaired" was "essentially synonymous" with maintaining natural conditions, a contention supported by interior secretary Franklin Lane's instruction to the first director of the National Park Service: "Every activity of the Service is subordinate to the duties imposed upon it to faithfully preserve the parks for posterity in essentially their natural state." Subsequent revisions of park management policy have further defined naturalness and made this concept the foundation of park stewardship (National

Park Service 2006). The Wilderness Act of 1964 also codified the centrality of naturalness as an attribute of wilderness character. Wilderness was established "to assure that an increasing population, accompanied by expanding settlement and growing mechanization, does not occupy and modify all areas . . . leaving no lands designated for preservation and protection in their natural condition."

The trouble with such heavy reliance on the concept of naturalness is that, like many terms, *natural* is a commonly used word with multiple meanings. When talking about what makes an area natural and how to keep it that way, different people use the term in very different ways and are often not conscious of how their definitions differ. For many people, naturalness implies a lack of human effect. Natural areas should be pristine, uninfluenced by humans, or at least modern technological humans. This means ensuring that the current composition, structure, and functioning of ecosystems are consistent with the conditions that would have prevailed in the absence of humans (either all humans or post-aboriginal ones) (Cole 2000). A place is natural if it is devoid of human artifacts and unaffected by such human threats and activities as pollution and fire suppression.

A related but distinctly different meaning of *naturalness* implies freedom from intentional human control. According to this meaning, an area may bear the mark of human presence, such as shelters and trails, and its ecosystems may have been altered by pollution, invasive species, and other threats; however, it can be a natural area if it is not subject to intentional manipulation and human intervention. In such a place, nature is self-willed, autonomous (Ridder 2007), left to its own devices, and free from the constraints of human intentionality.

Yet another meaning of *naturalness* implies a connection to the past. A natural area is one that is true to the historical condition of the ecosystem. Natural ecosystems should appear and function as they did in the past. This notion of historical fidelity is rooted both in a nostalgic connection to history and in an ethical duty to pass on to the future what was inherited from the past.

When managers face decisions such as those explored in Chapter 1—whether to cut trees in Bandelier or helicopter limestone into the Saint Mary's River to preserve ecological values—the diverse meanings of *naturalness* become entangled. In this chapter we trace how these meanings have evolved, from a time when they were considered to be congruent (by most protected area managers, at least) to the present day, when they are increasingly in conflict. We assert that changes in science and society and the globalization of human influence have eroded the adequacy of naturalness as a guiding concept for protected area stewardship, the thesis that will set the stage for the remainder of this book.

Early Meanings of *Naturalness*

Any word worth using must have meaning to both the user and the audience. Clearly, by its frequent use, *naturalness*, or more commonly its adjective form, *natural*, meets that test. Every day, people encounter not just natural areas but natural foods, natural athletes, natural gas, and natural history. But what do people really mean when they use the term?

The *American Heritage Dictionary* defines *natural* as "present in or produced by nature; not artificial or man-made," with the etymology rooted in the Latin *natura*, meaning "nature or birth." The same dictionary also provides a relevant definition of *nature* as "the physical world, usually the outdoors, including all living things." Although humans are clearly among "all living things," the traditional counterposition of natural and artificial is an ancient concept and has given rise to a dualistic separation of humans and nature. Even as nature, or the nonhuman world, came to be revered and valued by people, it was viewed as a divine "Other," godlike in its separation from humans (Cronon 1995).

As conservation caught on in the late nineteenth century, nature underwent a transition from Cronon's metaphysical "Other" to very real objects of loss. With the 1854 publication of George Perkins Marsh's *Man and Nature*, the human role in the disappearance of the natural world was formally acknowledged. As Oelschlager (1991: 107) observes, "A careful reading of *Man and Nature* leaves one incredulous, since Marsh marshaled almost irrefragable evidence, spanning an enormous array of activities, that humankind was on balance a destabilizing environmental force whose impacts portended an uncertain future." In the decades that followed, the modern conservation movement was born. As reports returned from the "vanishing frontier" and painters such as Albert Bierstadt decried "The Last of the Buffalo," the first wilderness parks and forest reserves were created to protect at least some of the disappearing, nonhuman natural world.

Here, at the beginning of the twenty-first century, it is hard to imagine the state of natural resources at the end of the nineteenth century. The once inexhaustible bison herds were all but gone, as were waterfowl and a host of other game birds. Forests were being cut and sold without any regard for the future. Natural resource management did not even exist as a professional field. A century ago, the only way to halt the violence was to draw a line around a place and protect its objects from the commercial onslaught. Preservation of natural conditions was equated with protection from exploitation. For example, when the Ecological Society of America proposed that a national system of nature sanctuaries be established, Shelford (1933: 245) asserted that "remedial measures directed toward the return to a so-called equilibrium consist chiefly in allowing nature to take its course."

Bolstering this sense that maintaining natural conditions could be achieved solely through protection from resource extraction was the prevailing ecological paradigm of the day. Ecology first emerged as a field of study in the latter half of the nineteenth century, but within just a few years, a school of thought was so firmly entrenched that it would guide protected area management for the next half century or more. Climax theory (*sensu* Clements 1916) held that all vegetation was at, or was returning to, a fully developed climax stage of succession that was natural and characteristic of the region. All one needed to do to preserve natural and historical conditions was to avoid disturbances such as logging, grazing, fire, and insect outbreaks. So, for decades after *nature* took on its modern, conservation-oriented meaning, managers and policymakers assumed that nature could be sustained simply by protecting parks from disturbance.

Evolving Ecological Science

By the mid-twentieth century, however, a new set of challenges to nature preservation was being advanced, this time from within the National Park Service. The science of ecology had progressed dramatically, and by the 1930s wildlife biologists surveying the national parks saw "a world of wounds" (sensu Leopold 1953), unraveling ecosystems characterized by wildlife extinctions, feral livestock, and overgrazed plant communities (Sellars 1997). Park Service scientists, led by George Wright, argued that protection from disturbance was inadequate to preserve park values. They asserted that certain types of direct intervention (e.g., elk herd reduction, predator reintroduction) were necessary to preserve natural conditions, whereas other activities that had previously been considered benign (e.g., scenery management and fire, insect, and predator control) diminished naturalness. But this perspective was controversial. Many in the Park Service disagreed, countering that scenery management and fire, insect, and predator control were needed to "preserve natural conditions" in a world that was no longer "in balance" (Sellars 1997).

The controversy waxed and waned throughout the 1940s and 1950s until 1963, when a report authored by A. Starker Leopold and colleagues boldly affirmed the position of the park wildlife biologists. The Leopold report (Leopold et al. 1963: 4) famously asserted that the goal of national park management should be to maintain "biotic associations . . . as nearly as possible in the condition that prevailed when the area was first visited by the white man. A national park should represent a vignette of primitive America." The report recognized that "most biotic communities are in a constant state of change," and maintaining natural ecosystems entails maintaining their dynamics. In many cases, this would entail active management, including herd reduction, prescribed fire, and reintroduction of extirpated species.

Anticipating the advent of restoration ecology by two decades, the Leopold Commission argued, "So far we have not exercised much imagination or ingenuity in rebuilding damaged biotas. It will not be done by passive protection alone" (Leopold et al. 1963: 10). The language of the Leopold report suggests the authors saw little or no conflict between maintaining historical conditions and minimizing human effects on ecosystems. Maintaining biotic associations and restoring them, where damaged, would apparently accomplish both. Although they had no qualms about achieving the "maintenance of naturalness" through active intervention, the authors did express concern that doing so would interject artificiality. They were more concerned about the appearance of artificiality than with artificiality per se, however. Where park ecosystems are actively managed, they wrote, "observable artificiality in any form must be minimized and obscured . . . hidden from visitors insofar as possible" (Leopold et al. 1963: 6).

For those whose jobs depended on having a working definition of "protecting natural conditions," the Leopold report provided a foundation that lasted for many years: Restore the conditions that existed before people messed it up, but leave the smallest mark possible. Where conditions are perceived to have changed little since the arrival of white settlers, there is no need for intervention, and nature can be protected through a light touch. Where fire exclusion has altered fuel loads, exotic species have altered species composition, food webs, and vegetation structure, or where air pollution has altered soil or stream chemistry, the preservation of nature entails active intervention, but it should be done with as little "observable artificiality" as possible. This perspective gave rise to programs of natural fire use, feral animal control, and wildlife reintroduction that provided a comfortable foundation to national park management for decades.

The Emergence of Wilderness Values

At about the same time as the Leopold report attempted to clarify the purpose of national park management, the Wilderness Act (P.L. 88-577, 78 Stat. 890) became law. Its language stressed a meaning of naturalness that differed from the focus on historical biotic communities emphasized in the Leopold report. The founders of the wilderness movement, which began in the 1920s, saw agency promotion of recreational motoring and the resultant tendency "to barber and manicure wild America" as the single greatest threat to the protection of nature, and they advocated a new form of management with a much lighter touch. As Paul Sutter (2002: 14) notes in *Driven Wild*, his history of the early wilderness movement, "The founders of the Wilderness Society did see wilderness areas as places meant to preserve pristine nature . . . [but] . . . wilderness was as much about 'wildness,' the absence of human control, as it was about pristine ecological conditions." The primary definition of wilderness, from the Wilderness Act, is a place "where the earth and its community of life are untrammeled by man." *Untrammeled* is often misinterpreted to mean *undisturbed*; however, it is not a descriptor of the ecological condition of the land (Scott 2001). Synonymous with *unconfined*, *unfettered*, and *unrestrained*, *untrammeled* suggests freedom from human control more than lack of human effect (Cole 2000). When Howard Zahniser, author of the Wilderness Act, selected the word *untrammeled* to characterize the nature of wilderness, he intended that the law protect nature by keeping our hands off: "We must never forget, we are guardians, not gardeners" (Zahniser 1963, inside cover).

By the end of the 1960s, naturalness was being applied as a concept for guiding the stewardship of protected areas, whereas once it had been primarily a reason to establish such places. Naturalness had evolved from a simple notion of protection from development and exploitation to an elaborate set of meanings, including minimizing human effect and influence, preserving historical conditions, and minimizing control over nature. Still, it was widely perceived, at least among protected area managers, that these three meanings were essentially the same thing, that they were congruent and compatible.

Naturalness Challenged

Eventually, accumulating scientific evidence from numerous disciplines began to challenge the perception that eliminating human impacts, maintaining historical fidelity, and not "trammeling" the land could all be achieved simultaneously on the same piece of ground.

Influence of Indigenous Populations

Research into the effects of indigenous peoples on wildland ecosystems (Day 1953) deflated the myth of the pristine wilderness (Mann 2005). Although the magnitude of past human influence was variable (Vale 2002), many park and wilderness ecosystems had been profoundly affected by humans by the time Europeans first arrived in North America (Anderson 1996). Long histories of burning and hunting, in particular, shaped the natural landscape. For example, Kay (1995) argues that native hunters kept elk populations low in Yellowstone National Park. After the creation of the park and the displacement of indigenous populations, burgeoning elk populations adversely affected willow and aspen communities, a change that has had cascading effects on beaver and other species.

Perhaps less controversial is the notion that indigenous burning has shaped many ecosystems (Pyne 1997). Indigenous people burned for many reasons, in different frequencies, intensities, locations, and seasons. Consequently, the effects of burning were widespread, often profound and variable from place to place (Kilgore 1985). Among other effects, burning maintained forest openings, decreased tree density, changed species composition, and sustained valued species. In some protected areas, in North America and certainly in other parts of the world, indigenous populations still live on the land and still shape the landscape, often in ways that are considered desirable (Gillson and Willis 2004).

The fact that many ecosystems perceived as natural were, and continue to be, substantially shaped by human activity erodes the meaning of *natural* as free from human effect. On sites that have been highly affected by human activity for millennia, the distinction between natural and artificial becomes blurred. If the purpose of protected areas is to preserve natural conditions and yet there is no objectively determined condition that can be called natural, the very purpose of protected areas is called into question. In response, some have sought to define naturalness differently, distinguishing temporally between early human influence, when technologies were less sophisticated and less of a threat to nature, and more recent human influence. For example, Landres et al. (1998: 44) suggests that, in the context of wilderness, *naturalness* should be defined as "unaffected by contemporary (roughly from the time of European settlement on) anthropogenic influences." In Australia, legislation stipulates that it is only the effects of "modern technology" or "modern society" that are inconsistent with wilderness; aboriginal influences are consistent with wilderness (Prest 2008). In contrast, Hunter (1996) argues that *natural* should be interpreted as "without human influence," regardless of the group of people involved or the timing of their influence, even if it meant natural areas would no longer support historical conditions.

The Rise of Nonequilibrium Dynamics and the Conservation of Biodiversity

Just as evidence of indigenous influence was coming to light, ecologists were documenting the role of disturbance in shaping ecosystems, which forced a reexamination of the way nature works (White 1979; Pickett and White 1985). Rather than tending toward some primeval "natural" or climax state that existed before white settlement, ecosystems were discovered to be nonequilibrium systems, constantly changing, particularly in response to disturbance. As Sprugel (1991: 15) notes, "The notion of 'natural' vegetation or ecosystem processes . . . must be revised to recognize that there is a range of ecosystems that can legitimately be considered." There is no single "natural condition" toward which the system would tend if left alone.

The significance of this nonequilibrium paradigm shift in ecology (Pickett et al. 1992; Fiedler et al. 1997) to protected area managers cannot be overstated. This shift raised difficult questions. What is the target for an intervention designed to restore naturalness to impaired park and wilderness ecosystems if there is no single natural condition? It also occurred at a time when the failure of traditional land management practices to sustain ecosystems, or at least ecosystem elements such as Kirtland's warbler (*Dendroica kirtlandii*), was becoming acutely obvious. Recognition that so many species depended on disturbance to maintain their habitat gave rise to the idea that ecosystems that managed to sustain their characteristic disturbance regimes stood a better chance of sustaining all their parts (Pickett and White 1985).

The timing of the nonequilibrium paradigm shift also corresponded with global recognition of a biodiversity crisis (Brundtland 1987). Conservation biologists documented the greatest rate of extinctions since the end of the Pleistocene. Protected areas took on renewed importance as a bulwark of reserves necessary to combat the global extinction crisis. The

biotic associations Leopold described as the objects of protected area management were understood to encompass the entire diversity of life, further
expanding the meaning of protecting naturalness. Indeed, the International
Union for the Conservation of Nature (2008: 13) defines a protected area
as "an area of land and/or sea especially dedicated to the protection and
maintenance of biological diversity, and of natural and associated cultural
resources, and managed through legal or other effective means."

Recognition of the importance of disturbance dynamics to the conservation of nature and the need for new approaches to the conservation
of biodiversity led to development of the concept of historical range of
variability (HRV) (Morgan et al. 1994), the idea that ecosystem characteristics are variable over time, but within bounds, and that maintaining them
within those bounds necessarily maintains their components. As Pickett et
al. (1992: 82) observe, "Human-generated changes must be constrained
because nature has functional, historical, and evolutionary limits. Nature
has a range of ways to be, but there is a limit to those ways, and therefore
human changes must be within those limits." The HRV concept amplified
the importance of disturbance found in the Leopold report but described it
in terms of bounded behavior that would sustain the historical condition of
the ecosystem. As Aplet (1999: 355) asserts, "It is the bounded condition
of ecosystems, dynamic and in the presence of aboriginal man, that we may
consider 'natural' or 'pristine.' "

By this definition, naturalness could be measured as the degree to
which a place retains the ecological composition and structure—dynamic
yet bounded over time—that characterized the system before the dramatic
anthropogenic modifications of the recent past. The more native species
and fewer exotic species a place retains, the more the patches retain the
character of those produced by the historical disturbance regime, and the
more consistent soil, air, and water quality are with historical ecosystem dynamics, the more natural the place is. Unfortunately, accumulating effects
of a century or more of grazing, logging, water diversion, fire suppression,
and species invasion have left many areas functioning outside HRV, and climate change threatens further alterations. Maintaining HRV increasingly
entails human intervention.

The Dilemma of Wilderness Management

With recognition that maintaining ecosystem composition and function
increasingly entails asserting human control, the formerly congruent mean-

ings of naturalness are conceptually split. The uneasy balance recommended by the Leopold report, which has characterized protected area management since the mid-twentieth century, cannot be sustained: In altered ecosystems, neither historical fidelity nor lack of human effect can be achieved without human control. Maintaining historical ecosystems or keeping ecosystems on the trajectory they would be on in the absence of human effects entails intentional and repeated human intervention. Cole (1996) calls this the "dilemma of wilderness management."

To illustrate this tension graphically, Aplet (1999) proposes a conceptual model (Figure 2.1) in which an axis defined by degree of manipulation or control (from controlled to "self-willed") is orthogonal to an axis describing ecological condition (from novel to pristine). For our purposes, this axis may represent either historical fidelity or lack of human effect. In the upper right corner of the figure occur the most uncontrolled, unaffected places, the large, ecologically intact landscapes where historical fidelity has been maintained without much human intervention. The Arctic National Wildlife Refuge is a prime example. Its antipode, the highly altered, highly controlled environment of the city, occurs in the lower left corner. Still other landscapes, such as the historically accurate but highly manipulated prairie restoration project at the University of Wisconsin Arboretum, belongs in the lower right-hand corner, and the C&O Canal, an

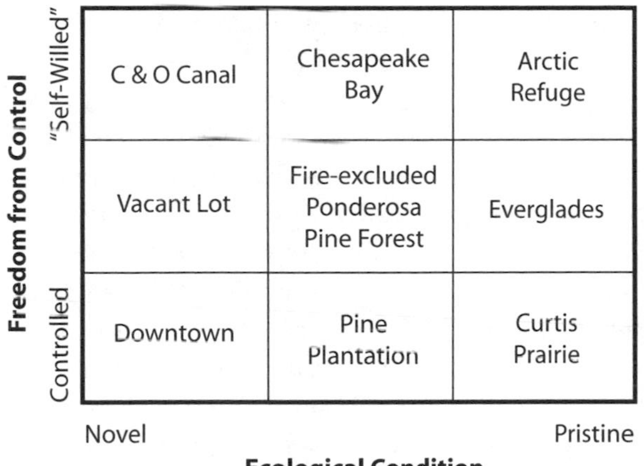

"Self-Willed"	C & O Canal	Chesapeake Bay	Arctic Refuge
	Vacant Lot	Fire-excluded Ponderosa Pine Forest	Everglades
Controlled	Downtown	Pine Plantation	Curtis Prairie
	Novel		Pristine

Freedom from Control

Ecological Condition

FIGURE 2.1. A conceptual model that arrays landscapes along two axes, from controlled to self-willed and from novel to pristine. The qualities these axes represent are consistent with traditional definitions of naturalness. Their use clarifies the difference in meaning between freedom from intentional human control and maintenance of historical or undisturbed conditions. (Adapted from Aplet 1999)

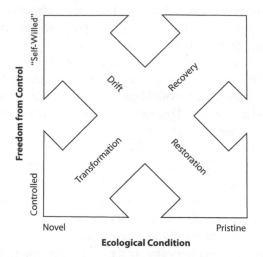

FIGURE 2.2. Like landscapes, stewardship options can be arrayed along two axes, from controlled to self-willed and from novel to pristine. (Adapted from Aplet 1999)

artificially constructed waterway parallel to the Potomac River, overgrown with exotic species, might reasonably be called ahistorical, highly altered, yet self-willed and untrammeled. Landscapes can express any combination of human control and historical fidelity. Where they are most untrammeled and unaltered, they are called wilderness.

This conceptualization can be used to contemplate the role of human agency in shaping landscape character (Figure 2.2) (Aplet 1999). Increased human effort can drive systems away from pristine conditions through transformation, as has typified the progress of civilization. This is clearly at odds with the definition of wilderness and national parks. Human effort can also be exerted to increase historical fidelity and mitigate human impacts, through the process of restoration. In the absence of active management, land freed from human control can either recover toward the pristine or drift toward a more novel condition. Franklin and Aplet (2002) assert that, for wilderness, recovery is always the ideal trajectory; however, they recognize that there will be cases in which recovery is impossible without active restoration. In these cases, the decision to intervene "will hinge on whether the potential for [recovery] outweighs the ecological uncertainties and the magnitude and duration of the required trammeling" (Franklin and Aplet 2002: 278).

The explicit separation of ecological condition from human control helped establish some precision in managing for naturalness. No longer was it appropriate to balance competing meanings of naturalness. Rather, it was important to be clear about which meaning to emphasize. This provided a frame for protected area managers to understand that their work requires choices, not balance. Protected area managers must increasingly "decide which of the two aspects of naturalness—pristine conditions or unmanipulated conditions—should be given preeminence" (Cole 1996: 16).

Ubiquitous and Directional Human Change

As appreciation of the implications of global environmental change increases, so does the need for managers to make choices, based on an understanding of differences between the three meanings of naturalness. Even the most remote places on Earth are affected by human activities (Vitousek et al. 2000). Every acre of every park and wilderness has been and will continue to be affected, to some degree, by the activities of modern technological humans. The major drivers of ecosystems are changing under the onslaught of invasive species, climate change, and other stressors. Sustaining historical ecosystem composition, structure, and function is increasingly difficult, even under heavy-handed human control. Many of these human-caused changes are not cyclical. Park conditions are not in dynamic equilibrium, varying around some functional, historical steady state. The drivers of change are directional. Future conditions will be very different from current conditions, perhaps well outside the bounds of historical variability.

Climate change, in particular, exposes the limitations of the naturalness concept. Paleoecological research (e.g., Pierce et al. 2004) reveals that fire frequency and severity have changed with shifts in climate in the past, and the increased fire activity that has characterized recent decades in the western United States (Westerling et al. 2006) can be expected to continue or worsen in the future. Species invasions have changed the rules of ecosystem dynamics over broad areas; for example, cheatgrass (*Bromus tectorum*) has permanently altered fire regimes of the Great Basin. Climate change means that park and wilderness ecosystems will inevitably be substantially affected by humans. That conclusion cannot be denied. The fact that climate change is directional and will have novel and unpredictable effects points out the limitations of HRV as a guide for the future—as a proxy of sustainable conditions. Restoration of past conditions may be a recipe for disaster if

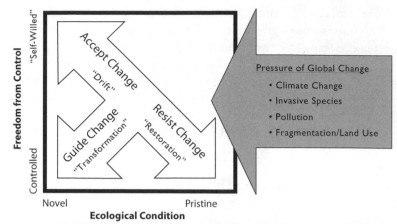

FIGURE 2.3. Global environmental change precludes the ideal stewardship option in parks and wilderness: that release from human control will increase historical fidelity and pristineness. Protected area managers must choose to increase historical fidelity through restoration, accept the change that will result from less intervention and control, or transform ecosystems to future states that are not true to the past but will protect important values and be more resilient in the face of global change.

the climatic conditions of the future are unfavorable for those ecosystems (Harris et al. 2006).

To reflect this, we have modified the model presented in Figure 2.2. Global environmental change represents an enormous pressure bearing down on the land, driving it away from historical and pristine conditions (Figure 2.3). As past controls on ecosystems change, so will those ecosystems. In the face of such pressure, recovery of historical conditions, unaided by human intervention, is not likely to be effective, even where human influence was historically minimal; consequently, the ideal approach to park and wilderness management is lost. Conservation of the species and ecosystems inherited from the past now depends on actively resisting change through restoration. Where land is untrammeled and intervention has been avoided, change is inevitable. Resultant ecosystems can no longer be expected to retain their historical character; rather, they will probably drift into new, unprecedented conditions, with unknown consequences for biodiversity. In this context, managers can either accept change or seek to guide change, using interventions to transform ecosystems into conditions more resilient to future climates, better able to conserve important ecological values. The tensions between resisting change, guiding change, and accepting change will be explored more fully in the second part of this book.

Complex and Conflicting Values

It is increasingly clear that a congruent perspective about what naturalness means, widely shared by the public, ecologists, and protected area managers, does not exist. As myths about natural systems have been deflated, the value of naturalness as a conceptual foundation on which to base operational management decisions has been called into question. If the ecosystems of parks and wilderness are unavoidably affected by humans, if the conditions of the past cannot—or even should not—be preserved, and if "untrammeling" does not lead to recovery, what does *natural* mean?

Stewardship of parks and wilderness has also been complicated by the ever-increasing array of values that have been attached to protected areas. Initially, national parks were largely about scenery and spectacle (Graber 1983), preserving favored biological and physical objects and keeping things the way they were. The advent of wilderness brought protection of places untrammeled by humans. In the late twentieth century, the conservation of biological diversity emerged as a core goal, with the definition of biological diversity expanding to include preservation of genetic diversity, species, plant and animal communities, and the fundamental physical and biological processes on which organisms depend (National Park Service 2006). Recently, there has been increased emphasis on the need for parks to provide opportunities for social engagement, for example, in restoration efforts designed to foster healthier human–nature relationships (Higgs 2003).

As park purposes and values expand, conflict between them inevitably increases. The goals of conserving all biodiversity and allowing unfettered evolution may conflict with the goals of protecting particularly valued species and preserving some places as they were in the past. Public participation in restoration can conflict with efforts to minimize human influence (Throop and Purdom 2006). Respect for nature's autonomy can conflict with the active efforts to reverse some of the deleterious effects of human activity on park and wilderness ecosystems. Managing for "naturalness" clearly does not help to resolve any of these conflicts.

Beyond Naturalness

In his provocative essay "The Trouble with Wilderness: Or Getting Back to the Wrong Nature," William Cronon (1995) argues that wilderness is not what it seems—that the unexamined and mythic meanings attached to wilderness divert attention and prevent realization of many important environmental values. He asserts that it is time to rethink wilderness. In particular, he argues for a need to move beyond couching our conception of wilderness and the values it embodies so firmly in a dualistic vision of humans apart from nature. In a similar vein, we assert that it is time to rethink naturalness, lest the unexamined, mythical, diverse, and conflicting meanings of the term prevent realization of many important purposes and values of parks and wilderness. In particular, it is time to articulate goals and objectives for parks and wilderness that are founded in a perspective that views humans as part of, rather than apart from, nature.

Since key enabling acts and management policies were established for national parks and wilderness areas, values, beliefs, and the world itself have changed. The conservation of biological diversity has become a core value of protected areas. Beliefs about the stability of ecological systems, the insignificance of aboriginal humans as ecological agents, and our ability to mitigate the adverse effects of current and future human activity on park ecosystems have all been shaken by research in ecology, paleoecology, anthropology, and related fields. The goals that guided the conservation and restoration of large protected areas in the twentieth century—most notably the concept of naturalness—do not provide sufficient guidance for future park and wilderness stewardship.

We began this chapter by noting that naturalness is a touchstone for many people. Despite conflicting and ambiguous meanings, the concept of naturalness continues to have value, even in the context of protected area stewardship. The notion of naturalness embodies society's interest in conserving the nonhuman elements of our world. The term has mobilized public concern and interest in park and wilderness protection, and it continues to provide an idealized, general vision of what preservation is intended to achieve. The concept of naturalness is not likely to disappear from the policy frameworks that guide protected area stewardship. But times have changed. As conservation imperatives have expanded beyond the setting aside of parks and wilderness areas to working within them to protect their values, new concepts are needed to guide management—concepts that can be drawn on to articulate a desirable and attainable future for park and wilderness ecosystems that accounts for human impacts, global change, and evolving public values.

In order to develop practical operational objectives to guide stewardship, it is essential to clarify park and wilderness purposes and values, indicating more specifically what should be sustained into the future. New guiding concepts must be advanced and considered. Innovations in management strategies, planning processes, and institutions must be experimented with.

Armed with an expanded array of tools and concepts, framed in more clearly articulated policy, managers are more likely to make stewardship decisions that will secure the values of parks and wilderness in perpetuity.

REFERENCES

Anderson, M. K. 1996. Tending the wilderness. *Restoration and Management Notes* 14:154–166.

Aplet, G. H. 1999. On the nature of wildness: Exploring what wilderness really protects. *Denver University Law Review* 76:347–367.

Brundtland, G. H. 1987. *Our common future: The report of the World Commission on Environment and Development*. Oxford University Press, Oxford.

Clements, F. E. 1916. *Plant succession: An analysis of the development of vegetation*. Publication No. 242. Carnegie Institute, Washington, DC.

Cole, D. N. 1996. Ecological manipulation in wilderness: An emerging management dilemma. *International Journal of Wilderness* 2(1):15–19.

Cole, D. N. 2000. Paradox of the primeval: Ecological restoration in wilderness. *Ecological Restoration* 18:77–86.

Cronon, W. 1995. The trouble with wilderness: Or getting back to the wrong nature. Pp. 69–90 in W. Cronon, ed. *Uncommon ground: Toward reinventing nature*. Norton, New York.

Day, G. M. 1953. The Indian as an ecological factor in the northeastern forest. *Ecology* 34:329–346.

Fiedler, P. L., P. S. White, and R. A. Leidy. 1997. The paradigm shift in ecology and its implications for conservation. Pp. 83–92 in S. T. A. Pickett, R. S. Ostfeld, M. Shachak, and G. E. Likens, eds. *The ecological basis of conservation: Heterogeneity, ecosystems, and biodiversity*. Chapman and Hall, New York.

Franklin, J. F., and G. H. Aplet. 2002. Wilderness ecosystems. Pp. 263–285 in J. C. Hendee and C. P. Dawson, eds. *Wilderness management: Stewardship and protection of resources and values*. Fulcrum, Golden, CO.

Gillson, L., and K. J. Willis. 2004. "As Earth's testimonies tell": Wilderness conservation in a changing world. *Ecology Letters* 7:990–998.

Graber, D. M. 1983. Rationalizing management of natural areas in national parks. *The George Wright Forum* 3:48–56.

Harris, J. A., R. J. Hobbs, E. Higgs, and J. Aronson. 2006. Ecological restoration and global climate change. *Restoration Ecology* 14:170–176.

Higgs, E. 2003. *Nature by design: People, natural process, and ecological restoration*. The MIT Press, Cambridge, MA.

Hunter, M. Jr. 1996. Benchmarks for managing ecosystems: Are human activities natural? *Conservation Biology* 10:695–697.

International Union for the Conservation of Nature. 2008. *Guidelines for applying protected area management categories*. IUCN, Gland, Switzerland.

Kay, C. E. 1995. Aboriginal overkill and native burning: Implications for modern ecosystem management. *Western Journal of Applied Forestry* 10:121–126.

Kilgore, B. M. 1985. The role of fire in wilderness: A state-of-knowledge review. Pp. 70–103 in R. C. Lucas, comp. *Proceedings, National Wilderness Research Conference: Issues, state-of-knowledge, future directions.* General Technical Report INT-220. USDA Forest Service, Intermountain Research Station, Ogden, UT.

Landres, P. B., P. S. White, G. Aplet, and A. Zimmermann. 1998. Naturalness and natural variability: Definitions, concepts and strategies for wilderness management. Pp. 41–52 in D. L. Kulhavy and M. H. Legg, eds. *Wilderness and natural areas in eastern North America.* Center for Applied Studies in Forestry, Stephen F. Austin State University, Nacogdoches, TX.

Leopold, A. 1953. *Round River.* Oxford University Press, New York.

Leopold, A. S., S. A. Cain, D. M. Cottam, I. N. Gabrielson, and T. L. Kimball. 1963. Wildlife management in the national parks. *Transactions of the North American Wildlife and Natural Resources Conference* 28:28–45.

Mann, C. C. 2005. *1491: New revelations of the Americas before Columbus.* Knopf, New York.

Marsh, G. P. 1864. *Man and nature; or, physical geography as modified by human action.* Charles Scribner, New York.

Morgan, P., G. H. Aplet, J. B. Haufler, H. C. Humphries, M. M. Moore, and W. D. Wilson. 1994. Historical range of variability: A useful tool for evaluating ecosystem change. *Journal of Sustainable Forestry* 2:87–111.

National Park Service. 2006. *Management policies 2006.* Retrieved October 12, 2009 from www.nps.gov/policy/MP2006.pdf.

Oelschlager, M. 1991. *The idea of wilderness: From prehistory to the age of ecology.* Yale University Press, New Haven, CT.

Pickett, S. T. A., V. T. Parker, and P. L. Fiedler. 1992. The new paradigm in ecology: Implications for conservation biology above the species level. Pp. 65–88 in P. L. Fiedler and S. K. Jain, eds. *Conservation biology: The theory and practice of nature conservation, preservation, and management.* Chapman and Hall, New York.

Pickett, S. T. A., and P. S. White. 1985. *The ecology of natural disturbance and patch dynamics.* Academic Press, Orlando, FL.

Pierce, J. L., G. A. Maycr, and A. J. T. Jull. 2004. Fire-induced erosion and millennial-scale climate change in northern ponderosa pine forests. *Nature* 432:87–90.

Prest, J. 2008. Australia. Pp. 56–89 in C. F. Kormos, ed. *A handbook on international wilderness law and policy.* Fulcrum, Golden, CO.

Pyne, S. J. 1997. *Fire in America.* University of Washington Press, Seattle.

Ridder, B. 2007. The naturalness versus wildness debate: Ambiguity, inconsistency, and unattainable objectivity. *Restoration Ecology* 15:8–12.

Scott, D. W. 2001. "Untrammeled," "wilderness character," and the challenges of wilderness preservation. *Wild Earth* 11:72–79.

Sellars, R. W. 1997. *Preserving nature in the national parks: A history*. Yale University Press, New Haven, CT.

Shelford, V. E. 1933. Ecological Society of America: A nature sanctuary plan unanimously adopted by the Society, December 28, 1932. *Ecology* 14:240–245.

Sprugel, D. 1991. Disturbance, equilibrium, and environmental variability: What is "natural" vegetation in a changing environment? *Biological Conservation* 58:1–18.

Sutter, P. S. 2002. *Driven wild: How the fight against automobiles launched the modern wilderness movement*. University of Washington Press, Seattle.

Throop, W., and R. Purdom. 2006. Wilderness restoration: The paradox of public participation. *Restoration Ecology* 14:493–499.

Vale, T. R. 2002. The pre-European landscape of the United States: Pristine or humanized? Pp. 1–39 in T. R. Vale, ed. *Fire, native peoples, and the natural landscape*. Island Press, Washington, DC.

Vitousek, P. M., J. D. Aber, C. L. Goodale, and G. H. Aplet. 2000. Global change and wilderness science. Pp. 5–9 in S. F. McCool, D. N. Cole, W. T. Borrie, and J. O'Loughlin, comps. *Wilderness science in a time of change conference, Volume 1: Changing perspectives and future directions*. Proceedings RMRS-P-15-VOL-1. USDA Forest Service, Rocky Mountain Research Station, Ogden, UT.

Westerling, A. L., II. G. Hidalgo, D. R. Cayan, and T. W. Swetnam. 2006. Warming and earlier spring increases western U.S. forest wildfire activity. *Science* 313:940–943.

White, P. S. 1979. Pattern, process, and natural disturbance in vegetation. *Botanical Review* 45:229–299.

Zahniser, H. 1963. Guardians not gardeners. *The Living Wilderness* 83:2.

PART I

The Changing Context of Park and Wilderness Stewardship

As Chapter 1 in this book described, human activities increasingly degrade parks and wilderness, compromising the very resources and values these lands were designated to protect. Anthropogenic change is increasing in both extent and magnitude. The managers responsible for stewarding parks and wilderness areas must decide whether they should respond to such changes by intervening in ecosystem processes. However, in thinking this through, park and wilderness managers often find they are damned if they do intervene and damned if they don't. By intervening, they interject artificiality into systems where the artificial is anathema. But if they do not intervene, important values, such as native biodiversity, ecosystem function, and ecosystem services, could be lost irretrievably. Where managers do intervene, they must decide what to do and be able to articulate the desired outcomes of their actions.

Naturalness is the central guiding concept in park and wilderness law and policy, the basis for deciding where, when, and how to intervene in biological and physical processes. Chapter 2 explored the varied meanings of *naturalness* and how they have evolved over time. It argued that, where once these meanings were considered by most protected area managers to be largely congruent, now it is clear that they are not. Managers cannot simultaneously, in the same place, protect and preserve all the park and wilderness values wrapped up in the concept of naturalness. Consequently, although preserving naturalness continues to be a useful way to articulate why we have parks and wilderness areas, the concept no longer provides a sufficient foundation for making difficult decisions about how we go about the business of preservation.

The following chapters describe three reasons for this evolution in thinking about naturalness. First, ecological knowledge has improved. Now it is clear that aboriginal populations had profound effects on many so-called natural ecosystems and that ecosystems are highly dynamic. Second, the reach of human impact on the globe has spread and the global nature of environmental change is better recognized. There is no place on the face of the earth—even in the most remote parts of the wilderness—that is not being and will not continue to be altered by modern human activity. Moreover, the drivers of change are directional and novel, suggesting that future conditions will be unprecedented and unpredictable to a substantial degree. Third, the purposes and values of parks and protected areas have diversified. They are complex and increasingly in conflict with each other.

In addition to exploring these challenges to naturalness, each of the following chapters provides insights that point to the approaches and strategies outlined later in this book. Chapter 3, on evolving ecological understandings, discusses some of the implications of a new ecological paradigm for protected area management. Of particular note is the realization that there is no single right endpoint that constitutes the natural state. Rather, there are many possible desirable future states, and managers must choose between them, even if they decide not to intervene. Ever-improving ecological knowledge can assist managers in making choices, devising interventions, evaluating consequences, and learning by doing. Chapter 4, on global climate change, other anthropogenic stressors, and their ecological effects, emphasizes the uncertain future for park and wilderness ecosystems. This uncertainty argues for pursuing diverse management goals and strategies, reducing risk by not putting all one's eggs in the same basket, and developing institutions and planning procedures that are better adapted than current ones to managing in the face of uncertainty. Chapter 5, on changing practices and policies, draws attention to the challenges of translating general and sometimes conflicting federal policies into specific management practices in particular parks or wilderness areas. Because current policies provide little specific direction regarding whether, when, and how to intervene and what the goals of such interventions should be, managers struggle when deciding what course of action is appropriate. Management discretion results in diverse approaches that, because they are unplanned and the result of happenstance, do not achieve the benefits of purposeful, large-scale planning, such as learning and spreading risk.

Together these chapters make the case that we need to move beyond naturalness to consider new goals and approaches for protected area management. They illuminate the problems and challenges that later authors seek to address.

Chapter 3

Evolving Ecological Understandings: The Implications of Ecosystem Dynamics

RICHARD J. HOBBS, ERIKA S. ZAVALETA, DAVID N. COLE, AND PETER S. WHITE

> But if getting older has taught me anything it is the paradoxical certainty
> that one shouldn't be certain about anything.
>
> —*Stephanie Dowrick*

The science of ecology and the movement to designate national parks and other protected areas began in the nineteenth century, maturing and evolving over the last 100 years. Ecological thinking changed greatly in the last century, and new theories about how ecosystems work have profound implications for the management of protected areas. In particular, an improved understanding of the dynamic nature of ecosystems suggests important changes to the ways protected area managers think about ecological values at risk, the need for management intervention in ecosystem processes, the desired outcomes of those interventions, and the specific techniques used to achieve objectives. In particular, as discussed in Chapter 2, park stewardship continues to be guided to a substantial degree by naturalness, a concept that appears less useful through the lens of modern ecology.

Although initially focused primarily on individual species and ecological phenomena in isolation, ecologists increasingly study systems, all species, their interactions, and the external and internal processes that drive ecosystems. Ecologists have expanded the spatial and temporal scales of phenomena they examine. By studying ecosystems at different scales, researchers have gained new insights about how such systems are organized, how they operate, what drives them, and how they might be managed. Ecologists increasingly understand that the structure and function of ecosystems reflect complex interactions of environmental and biotic processes at many scales, including longer-term and larger scales than the age and extent of parks. Analogous to this expanded ecological focus, notions of conservation have evolved to embrace a broad range of biodiversity and ecological processes, both of which are now seen as key values of protected areas.

Arguably the most important advance in ecological understanding, in terms of managing protected areas, has been recognition of the inherent dynamism of ecosystems and the importance of disturbance processes. A number of authors have written about a paradigm shift in ecology, from the old "balance of nature" perspective to the modern perspective of nature in flux. Pickett and Ostfield (1995: 274) describe the old paradigm as one that "assumed that ecological systems were closed, self-regulating, possessed of single equilibrium points reached by deterministic dynamics, rarely disturbed naturally, and separate from humans." Today, ecologists recognize that ecological systems are "open, regulated by events arising outside of their boundaries, lacking or prevented from attaining a stable point equilibrium, affected by natural disturbance, and incorporating humans and their effects" (Pickett and Ostfield 1995: 275). The fact that ecological systems are complex, dynamic, chaotic, and inevitably affected by human activity increases the challenge of protected area stewardship and exposes some of the contradictions inherent in the varied meanings of naturalness, the traditional goal of protected area stewardship.

In this chapter, we outline recent advances in ecological thinking that revise or challenge earlier understandings of ecosystems. We offer suggestions about what these advances mean for protected area management, particularly implications for the traditional concept of naturalness, alternative goals for ecosystem interventions, and various management practices. We begin by discussing the rise of biodiversity conservation as a central goal of park management. The remainder of the chapter explores the dynamism of ecosystems. As the pace of biophysical and sociocultural change accelerates, effective management more than ever requires clear enunciation of goals and strategies, design and implementation of suitable interventions, and the capacity to monitor key ecosystem attributes to detect change and determine whether interventions were successful.

Biodiversity and Conservation Goals

Protected areas are increasingly viewed as strongholds and refuges for native biological diversity (biodiversity). But the meaning of biodiversity conservation has expanded along with the field of ecology. Sometimes biodiversity is defined narrowly as a list of native species. Other times it is defined so broadly as to lose its meaning, as when it refers to everything from ecological processes to aesthetics. The Convention on Biological Diversity defines biodiversity (or biological diversity) as "the variability among living organisms from all sources, including, inter alia, terrestrial, marine, and other aquatic ecosystems, and the ecological complexes of which they are part: this includes diversity within species, between species and of ecosystems" (Article 2, www.cbd.int/convention/articles.shtml?a = cbd-02). We view this definition as a useful start to explore what *native biodiversity*, in practice, can usefully mean.

The Earth Summit definition emphasizes the value of biodiversity at a range of scales, including genes, populations, species, patches, ecosystems, and landscapes. Part and parcel of this cross-scale definition is the recognition of the interdependence of different scales via the processes that maintain biodiversity (Noss 1990). For example, without genetic and population diversity, species persistence is compromised because of the loss of the variability that permits evolutionary adaptation. Ecosystem and landscape diversity depend on large regional species pools to provide the raw materials of succession and assembly of the biotic community in any given area, and the maintenance of regional species pools depends on ecosystem and landscape diversity to support a wide range of species through time. Therefore, much more than just species and the assemblage in any given location must be conserved.

In the definition of goals for biodiversity conservation, special emphasis has been placed on the species most critical to ecosystem function, which could include keystone species and ecosystem engineers (Mills et al. 1993; Wright and Jones 2006). For example, sea otters (*Enhydra lutris*) exert tremendous influence over the maintenance of whole kelp forest ecosystems by limiting populations of kelp-consuming urchins (Estes and Duggins 1995). Because of this keystone role, extensive conservation measures have been extended to sea otters, including legislation, enforcement, translocations, and applied research. However, it may not be possible to confidently define a limited set of ecosystem functions that are important or to determine whether the species that contribute to these functions and features of structure (not to mention others inadvertently left out) in one place and time are sufficient over larger temporal and spatial scales. The "insurance hypothesis" suggests that retaining all species, even rare ones, in a system provides resilience in the face of changing environmental conditions (Yachi and Loreau 1999). For instance, a recent long-term study in a California grassland showed that a plant species that was rare for 20 years made up a significant proportion of the plant community in years with a particular climatic signature (Hobbs, Yates, and Mooney 2007). Thus, a rare species played a more dominant role during times of drought.

Finally, the concept of ecosystem services has recently grown in popularity and been linked to biodiversity, with implications for how to manage and what to protect. Through the way they function, ecosystems provide services that are essential for human well-being and survival, such as provision of water, food, and fiber, aesthetic beauty, and recreation opportunities (Daily 1997). Conservation efforts targeting ecosystem services can shift priorities from protecting all the constituent parts of biodiversity toward protecting the landscapes, watersheds, or aspects of ecosystem structure and function that are most critical to the provision of ecosystem services. From a protected area perspective, the addition of ecosystem structure, functioning, and services broadens the set of goals for management and the set of reasons for protecting native biodiversity. It also makes protected area stewardship more challenging because the means of optimizing one goal may be detrimental to another goal. For example, desired fire regimes for maximizing forest carbon storage and old-growth–dependent species populations might differ from the optimal fire regimes for maintaining landscape-scale forest and habitat diversity. Moreover, this expanded set of goals means that managers need to be explicit about goals and their relative priority.

From the Balance to the Flux of Nature

Insights from paleoecology, long-term ecological studies, and ecological modeling have revealed that ecosystems are highly dynamic and that they change in complex, nonlinear, and unpredictable ways. The idea of the balance of nature has been replaced with the flux of nature (Lodge and Hamlin 2006; Pickett and Ostfield 1995). Ecosystems may occur in a number of alternative states, and those states are contingent on many variables, particularly the history of disturbance and human intervention. Given their dynamic nature, ecosystems inevitably change on a number of spatial and temporal scales.

The implications of this paradigm shift for the stewardship of parks and wilderness are profound (Christensen 1988). As threats to biodiversity mount, managers increasingly consider interventions for protected area ecosystems and processes. But the shifting nature of ecosystems suggests that there are many possible desirable outcomes for such interventions; there is no single true natural state. Moreover, the fluctuations themselves matter. The traditional notion of restoring past conditions is often inappropriate because many of the changes in ecosystem state that demand intervention are irreversible. So what criteria should be used to make decisions about where and how to intervene and to define desired outcomes? In the pages that follow, we explore some of what is known about ecosystem dynamics and the implications for such decisions.

Ecosystem Dynamics and Disturbance

Ecologists classify ecosystem dynamics as deterministic or stochastic or as transitions between alternative stable states (multiple equilibria) (Figure 3.1). Understanding how ecosystems actually change through time is essential to management decisions. Assuming one type of dynamics when, in fact, another type is in train may lead to potentially unpleasant surprises and unwanted outcomes. This is particularly true if slow, predictable dynamics are assumed and more stochastic or state-change dynamics are actually in play. Deterministic dynamics assume a predictable path of ecosystem development given any particular starting point, regardless of conditions before or during the development of the ecosystem. This deterministic paradigm is embodied by the traditional Clementsian perspective of succession, which sees plant community development as occurring in a very predictable, repeatable way depending on the local climate and soil type (Clements 1916). Odum (1969) later expanded these ideas to include a wide range of ecosystem properties that could be expected to change in a predictable way through time. In limited cases and contexts, ecosystems do undergo fairly deterministic succession. Old-field succession in eastern North America provides a classic example of predictable succession, where after abandonment the plant community is dominated first by annuals, then herbaceous perennials, then shrubs, and finally a progression of different types of trees (Hobbs and Walker 2006).

Most ecologists now agree that deterministic models are too simple to describe the complexity of dynamics actually observed in most situations and that stochastic (chance, unpredictable) processes play important roles

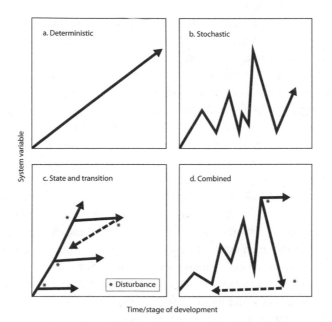

FIGURE 3.1. Ecosystem dynamics can be depicted as (a) deterministic, (b) stochastic, or (c) state and transition. In reality a combination of all three is likely (d). Traditionally, ecosystems have often been managed on the assumption that they are exhibiting deterministic dynamics. However, increasing levels of unexpected ecosystem change and surprise have led to a new understanding that incorporates the potential for both stochastic and state and transition dynamics. Solid lines represent progressive development, and dotted lines represent regressive dynamics. (Hobbs, Jentsch, and Temperton 2007)

(Temperton and Hobbs 2004). Although some apparently deterministic patterns are observable, there are usually variations around them, and often multiple developmental pathways are possible (Cramer 2006; Hobbs and Walker 2006). These variations often relate to the timing and severity of particular disturbances, climatic events, or soil- and resource-related phenomena. Abiotic disturbances in particular, such as fire, floods, hurricanes, and windthrow events, shape many (if not most) ecosystems. Altered disturbance regimes can profoundly change the structure, composition, and functioning of ecosystems. For example, the removal or reduction through dam building of pulse flood events from many rivers in the western United States has contributed to fundamental ecosystem changes such as widespread dominance by exotic tree species and large reductions in channel volumes (Stromberg and Chew 2002).

Managing disturbance regimes is often seen as a way to modify community assembly and succession. However, human disturbance regimes (such as altering grazing or fire regimes) must be tested for their historic bounds of variation. Disturbances that are within the historic bounds of variation for an ecosystem may produce different responses, or serve different management goals, than disturbances that are novel or create conditions that are outside those bounds. Altered fire regimes in western North American forests provide a vivid example of how ecosystems can change dramatically if historic disturbance regimes are altered (Arno and Fiedler 2005). Fire suppression during the twentieth century moved the forests away from their historic state and allowed a buildup of fuel that resulted in catastrophic wildfires in many areas. Stochastic dynamics suggest a less orderly development, with the path of development constantly being affected by a variety of factors, each working on a range of different temporal and spatial scales and resulting in an apparently random path.

Alternatively, the notion of transitions between alternative stable states suggests that an ecosystem can exist in a number of different configurations and can be driven from one state to another by particular disturbance, climate, or management events (Hobbs and Suding 2009; Suding et al. 2004; Suding and Gross 2006). In other words, development of an ecosystem is punctuated and characterized by sudden changes from one stable state to another. Such dynamics are being increasingly proposed to explain changes in ecosystems, originally in arid and freshwater systems but more recently across a wide range of different systems (Hobbs and Suding 2009). For example, the removal of perennial vegetation in arid regions can trigger positive feedbacks that lead rapidly to a stable sandy desert condition. Nutrient loading in lakes can produce little visible change until a threshold is crossed, rapidly rendering a clear lake turbid and dramatically changing its species composition (Scheffer et al. 2001).

Disturbance events of extreme magnitude, such as tsunamis or catastrophic fires, can cause deterministic and stochastic dynamics to be nonlinear in the sense that no near-term return to prior reference conditions or reference dynamics is possible. For example, landslides on Hawaii that remove ash-derived soils can alter forest successional trajectories and generate persistent changes, such as reduced biomass and increased exotic species dominance (Restrepo et al. 2003). State and transition dynamics also include the idea of thresholds where sudden changes in ecosystem properties occur.

It is thus increasingly clear that no one model of ecosystem dynamics is completely appropriate in all situations and for all systems (Cramer et al.

2008). As indicated in Figure 3.1, the same system may exhibit all three types of dynamics at different times, at different phases of development, or under different conditions. Assuming that one type of dynamics is in play when in fact another or a mix of dynamics is prevalent can lead to serious management problems. For example, assuming that a eutrophied lake can be restored simply by returning lake nutrient levels to pre-eutrophication levels could easily lead to a costly failure.

Thresholds and Alternative States

An important insight in ecology from the last 50 years is that nonlinear threshold changes can occur in ecological processes, species interactions, and population sizes in which there is a sudden switch from one state to a markedly different one (Walker et al. 2004). This switch can result from a particular event (such as a climatic extreme or major storm), management activity (such as the application of fire or changes in the grazing regime), or a nonlinear response to environmental change (such as pollution in a river reaching toxic concentrations). Crossing some thresholds may produce changes that are either irreversible or extremely difficult to reverse. For instance, too frequent fire can shift a shrub-dominated plant community to a herbaceous community that appears resistant to further change, particularly redevelopment of the previous shrub community (Keeley 2006). Consequently, understanding thresholds is critical to the development and evaluation of management actions.

Invasive Species and Rapid Evolution

Another factor affecting ecosystem dynamics and notions of naturalness is the increasing role that species not originally native to the locality have in ecosystems, as discussed in detail in Chapter 10. Today's conditions increasingly challenge managers to consider how to define and tackle invasive species. For example, where climate change is forcing species to migrate to new areas or other stressors have eliminated formerly dominant native species, nonnative species can meet important habitat needs for native species or perform useful ecosystem functions (Graves and Shapiro 2003). For instance, tamarisk invades riparian areas in western North America and alters ecosystem processes such as sedimentation, but some argue that, in the absence of native riparian tree species, it now provides habitat for threatened bird species (Cohn 2005). Therefore, efforts to remove tamarisk must be accompanied by the restoration of native riparian species. Where nonnative species have come to support native biodiversity in the absence of some now-missing species, ecosystem structure, or function, managers may decide to tolerate, retain, or even actively encourage clearly exotic species (such as Aldabran tortoises, *Geochelone gigantea*, in Mauritius) (Zavaleta et al. 2001). The pace and extent of change at scales far beyond individual protected areas challenge managers to think carefully about how to define *native* and *invasive* for practical purposes and about whether and when to accept the presence of a species even when it is recognized as an invader. The ecological and management implications of novel species assemblages have only recently begun to be considered seriously in the ecological literature (Hobbs et al. 2006; Seastadt et al. 2008).

As a final twist in how ecological systems are changing in ways that challenge traditional management approaches, there are increasing reports of species evolving rapidly in their new habitats (Thompson 1998), especially in the context of invasive species and pathogens (Strayer et al. 2006). Invasive species sometimes show a remarkable ability to evolve rapidly and adapt to their new habitat or biotic assemblage. A recent dramatic example of this is the Gough Island mouse (Wanless et al. 2007). The common house mouse *Mus musculus* was thought to have been inadvertently transported to the sub-Antarctic island by British whalers around 150 years ago. The mouse now appears to have evolved to two to three times the size of ordinary British house mice and from eating predominantly insects and seeds to eating the chicks of large birds, changes that may be causing problematic population declines. Therefore, it becomes increasingly important not to assume that a particular species will continue to behave in a similar manner to that found in its previous habitat. This implies the need for ongoing monitoring efforts to be robust enough to detect such changes and impacts.

The potential for rapid evolution is also an important factor in likely biotic responses to climate change. A range of species have already shown the capacity to undergo genetic change in the face of rapidly changing climates (Bradshaw and Holzapfel 2006), and yet this has rarely been factored into attempts to understand and predict biotic response to climate change.

Dynamic Ecosystems Challenge Notions of Naturalness

What do the recent developments in thinking about ecosystem dynamics mean to the concept of naturalness? The old "balance of nature" perspective on the world was consistent with notions of naturalness: Nature existed apart from humans and was largely predictable and static except when recovering from infrequent disturbance. These assumptions have slowly been pulled apart. Ecosystems are only seldom predictable and are in a constant state of dynamism, with individual species and biotic communities constantly changing in abundance through space and time. Furthermore, science has revealed that humans have been part of most ecosystems for millennia.

As discussed in Chapter 2, increased understanding of the widespread influence of indigenous humans on ecosystems around the world and the likelihood that indigenous people were keystone species exerting a strong influence on many environments necessitates rethinking how we view the past and plan for the future. If ecosystems evolved with human input, the absence of all human influence may not be desirable in some cases. For example, intentional burning could be necessary to maintain the open, park-like conditions of ponderosa pine forests and montane meadows in parts of the western United States. Instead, protected area managers may strive to minimize certain types of human effects. When managers determine whether a particular change is good or bad, they should no longer view human-induced change as necessarily bad, as implied by naturalness. Rather, criteria for what is good and bad should focus on various ecosystem attributes, such as species preserved, ecological integrity, historical fidelity, or resilience, as discussed in subsequent chapters. Another way of saying this is that the criteria for deciding whether to intervene and the desired outcomes of intervention should be shifted substantially from cause to effect. Whether or not the characteristics of the ecological system are acceptable or desirable is more important than whether or not change is human induced.

New ways of thinking about ecosystem dynamics lead further away from the old ideas about the balance of nature and the existence of one true "natural" state. If ecosystems are dynamic, the "natural state" is a moving target. What is natural may encompass many different potential states in any particular place though time and is likely to change as a result of climatic and disturbance events and changing management regimes.

Putting Ecosystem Dynamism into Practice

Ecosystem dynamism presents both challenges and opportunities for protected area managers. On one hand, confronting this level of complexity can be quite paralyzing because of the number of decisions to be made, the range of options that might be available, and the difficulty in monitoring simple cause-and-effect responses in an ecosystem. There may not be one correct state to aim for, and even without considering human effects on ecosystems, there is no single definable baseline. In addition, managers and society are faced with the possibility that certain introduced species might even be good (if they promote desirable goals and objectives).

Setting management goals is further complicated by the need to allow for natural disturbance processes. Historical range of variability (HRV) (Landres et al. 1999) can account for variation associated with disturbance (although directional change makes even this problematic). However, using HRV in a management context means that only conditions within the historic range are considered desirable. Applying HRV also entails arbitrary decisions about scale, place, and time. If scale, place, or time becomes too broad, almost anything is within the bounds of HRV, and the concept no longer provides practical guidance to the manager. Thus, maintaining historic species assemblages may not be the only valid management goal in protected areas. Novel species assemblages arising through species migration and invasion are increasingly inevitable as climate change and species movement increase; some of these novel assemblages might be desirable. Furthermore, as demonstrated by new ecological thinking about ecosystem dynamism, there are many possible desirable future states that are best considered in terms of ecosystem attributes (e.g., resilience, biodiversity) rather than degree of historic fidelity.

If the concept of naturalness is no longer a useful guide to developing goals for management interventions, what might replace or supplement it? Recognizing complexity in dynamics does not necessarily mean increased difficulty in deciding on management options. Despite the challenges described here, the dynamic nature of ecosystems may mean that a greater variety of valid management responses are available. In other words, one consequence of the new understanding of ecosystem dynamics is the expanded range of possible desirable outcomes. This means that there is less reason to seek the precise, correct future trajectory or endpoint. For example, intensive management to secure a particular mix of species and proportions of forest patches in a landscape could shift to a focus on maintaining disturbance dynamics and the ability for species to move across the landscape. It also means there is more need to make choices with little guidance regarding how to choose between alternatives. Managers will have to "play God" to a greater degree, and more frequently than they may want to. Because there are multiple ecosystem states and no clear correct choice, it is extremely important to articulate clear and specific goals.

Particularly given uncertainty, this suggests a need for diversity in the choices that are made, incorporating some built-in redundancy, planned at large spatial scales and implemented within an adaptive framework (see Chapters 12 and 13 for additional detail). Although there will probably be no one-size-fits-all approach that can be rolled out everywhere, it is possible to consider the suite of goals to be considered and the types of approaches needed (Lindenmayer et al. 2008). More attention must be paid to the process of prioritizing effort, using either analytical or triage-based approaches (Hobbs et al. 2003; Wilson et al. 2007). For example, Hobbs and colleagues (2003) outline an approach that prioritizes restoration actions based on the value of an ecological asset, the level of threat it faces, and the probability that the intervention will succeed. Rather than simple rule-based decision making inherited from past experience, new approaches will involve more decision tree analysis. When there is no one right answer, the range of possibilities and the likely consequences of action and inaction must be made as explicit as possible.

Applying the New Ecology

In this chapter we have suggested that ecosystems are dynamic and change in unpredictable ways, an insight that overturns traditional ways of thinking about ecosystems as stable, static, and balanced. These new ways of thinking about ecosystem flux and change reveal high levels of complexity and uncertainty, but the science of ecology can help managers deal with this complexity. This at first seems paradoxical. Indeed, it is true that dynamic ecosystems provide managers fewer certainties on which management prescriptions can be based. In the future, protected area managers will have to embrace the dynamism and uncertainty inherent in ecosystems, recognizing that many different future states are possible and potentially desirable. Given the unprecedented environmental changes of the twenty-first century, simply recognizing that the old rules do not necessarily apply is an important initial step toward a realistic view of how ecosystems work. It is critical to accept that ecosystems are not stable and predictable, tending toward historic conditions. Nor are they like machines that can be understood through isolated information about individual parts. Dynamic ecosystems are complex and uncertain, which necessitates trying different things in different places and responding to changing circumstances in an adaptive way. It involves keeping a close eye on how the system is behaving and making careful decisions about when and where to intervene. This is difficult but not impossible. Many of the "new" approaches demanded by our changed understanding of ecosystems are simply extensions of what we know should be good practice anyway. For instance, managers in Kruger National Park in South Africa have initiated a system of identifying "thresholds of potential concern" (Biggs and Rogers 2003), which act as triggers for particular types of intervention. This type of thinking moves away from the institutional tendency prevalent in many situations, which is to ignore potential problems until they become critical, only then instigating crisis management. It also demands the input of timely local information and expertise to inform decisions rather than reliance on broad general rules. In other words, the protected area manager needs to know his or her own turf and make decisions contingent on the local conditions but cognizant of the broader landscape and regional perspective.

Learning how to intervene effectively and efficiently in the dynamics of ecosystems is the key task for the coming decades. Moving on from naturalness demands that managers and scientists alike come to grips with the full extent of this task and find effective ways of making it work.

BOX 3.1. WHAT DYNAMIC ECOSYSTEMS MEAN FOR PROTECTED AREA MANAGEMENT

- Advances in ecological understanding over the past few decades challenge traditional notions of naturalness.
- It is now clear that ecosystems are not stable and predictable, tending toward historic conditions. This has profound implications for the stewardship of parks and wilderness.
- When managing ecosystems, it is important to embrace disturbance and different types of dynamism as key characteristics of ecosystems.
- When designing interventions, managers must understand that there is no single "correct" or "natural" ecosystem state to aim for. They should focus on outcomes and specific conservation goals rather than on whether change is caused by humans.

REFERENCES

Arno, S. F., and C. E. Fiedler. 2005. *Mimicking nature's fire*. Island Press, Washington, DC.

Biggs, H. C., and K. H. Rogers. 2003. An adaptive system to link science, monitoring and management in practice. Pp. 59–82 in J. du Toit, K. H. Rogers, and H. C. Biggs, eds. *The Kruger experience: Ecology and management of savanna heterogeneity*. Island Press, Washington, DC.

Bradshaw, W. F., and C. M. Holzapfel. 2006. Evolutionary response to rapid climate change. *Science* 312:1477–1478.

Christensen, N. L. 1988. Succession and natural disturbance: Paradigms, problems, and preservation of natural ecosystems. Pp. 62–86 in J. K. Agee and D. R. Johnson, eds. *Ecosystem management for parks and wilderness*. University of Washington Press, Seattle.

Clements, F. E. 1916. *Plant succession: An analysis of the development of vegetation*. Publication No. 242. Carnegie Institute, Washington, DC.

Cohn, J. P. 2005. Tiff over tamarisk: Can a nuisance be nice, too? *BioScience* 55:648–656.

Cramer, V. A. 2006. Old fields as complex systems: New concepts for describing the dynamics of abandoned farmland. Pp. 31–46 in V. A. Cramer and R. J. Hobbs, eds. *Old fields: Dynamics and restoration of abandoned farmland*. Island Press, Washington, DC.

Cramer, V. A., R. J. Hobbs, and R. J. Standish. 2008. What's new about old fields? Land abandonment and ecosystem assembly. *Trends in Ecology and Evolution* 23:104–112.

Daily, G. C., ed. 1997. *Nature's services: Societal dependence on natural ecosystems*. Island Press, Washington, DC.

Estes, J. A., and D. O. Duggins. 1995. Sea otters and kelp forests in Alaska: Generality and variation in a community ecological paradigm. *Ecological Monographs* 65:75–100.

Graves, S. D., and A. M. Shapiro. 2003. Exotics as host plants of the California butterfly fauna. *Biological Conservation* 110:413–433.

Hobbs, R. J., S. Arico, J. Aronson, J. S. Baron, P. Bridgewater, V. A. Cramer, P. R. Epstein, et al. 2006. Novel ecosystems: Theoretical and management aspects of the new ecological world order. *Global Ecology and Biogeography* 15:1–7.

Hobbs, R. J., V. A. Cramer, and L. J. Kristjanson. 2003. What happens if we can't fix it? Triage, palliative care and setting priorities in salinising landscapes. *Australian Journal of Botany* 51:647–653.

Hobbs, R. J., A. Jentsch, and V. M. Temperton. 2007. Restoration as a process of succession and assembly mediated by disturbance. Pp. 150–167 in L. R. Walker, J. Walker, and R. J. Hobbs, eds. *Linking restoration and ecological succession*. Springer, New York.

Hobbs, R. J., and K. N. Suding, eds. 2009. *New models for ecosystem dynamics and restoration*. Island Press, Washington, DC.

Hobbs, R. J., and L. Walker. 2006. Old field succession: Development of concepts. Pp. 17–30 in V. A. Cramer and R. J. Hobbs, eds. *Old fields: Dynamics and restoration of abandoned farmland*. Island Press, Washington, DC.

Hobbs, R. J., S. Yates, and H. A. Mooney. 2007. Long-term data reveal complex dynamics in grassland in relation to climate and disturbance. *Ecological Monographs* 77:545–568.

Keeley, J. E. 2006. South coast bioregion. Pp. 350–390 in N. G. Sugihara, J. W. van

Wagtendonk, K. E. Shaffer, J. Fites-Kaufman, and A. E. Thode, eds. *Fire in California's ecosystems*. University of California Press, Berkeley.

Landres, P. B., P. Morgan, and F. J. Swanson. 1999. Overview of the use of natural variability concepts in managing ecological systems. *Ecological Applications* 9:1179–1188.

Lindenmayer, D., R. J. Hobbs, R. Montague-Drake, J. Alexandra, A. Bennett, M. Burgman, P. Cale, et al. 2008. A checklist for ecological management of landscapes for conservation. *Ecology Letters* 11:78–91.

Lodge, D. M., and C. Hamlin, eds. 2006. *Religion and the new ecology: Environmental responsibility in a world in flux*. University of Notre Dame, Notre Dame, IN.

Mills, L. S., M. E. Soulé, and D. F. Doak. 1993. The keystone species concept in ecology and conservation. *BioScience* 43:219–224.

Noss, R. F. 1990. Indicators for monitoring biodiversity: A hierarchical approach. *Conservation Biology* 4:355–364.

Odum, E. P. 1969. The strategy of ecosystem development. *Science* 164:262–270.

Pickett, S. T. A., and R. S. Ostfield. 1995. The shifting paradigm in ecology. Pp. 261–278 in R. L. Knight and S. F. Bates, eds. *A new century for natural resources management*. Island Press, Washington, DC.

Restrepo, C., P. Vitousek, and P. Neville. 2003. Landslides significantly alter land cover and the distribution of biomass: An example from the Ninole ridges of Hawai'i. *Plant Ecology* 166:131–143.

Scheffer, M., S. Carpenter, J. A. Foley, C. Folke, and B. Walker. 2001. Catastrophic shifts in ecosystems. *Nature* 413:591–596.

Seastadt, T. R., R. J. Hobbs, and K. N. Suding. 2008. Management of novel ecosystems: Are novel approaches required? *Frontiers in Ecology and the Environment* 6:547–553.

Strayer, D. L., V. T. Eviner, J. M. Jeschke, and M. L. Pace. 2006. Understanding the long-term effects of species invasions. *Trends in Ecology and Evolution* 21:645–651.

Stromberg, J. C., and M. K. Chew. 2002. Flood pulses and restoration of riparian vegetation in the American Southwest. Pp. 11–49 in B. Middleton, ed. *Flood pulsing and wetlands: Restoring the natural hydrological balance*. Wiley, New York.

Suding, K. N., and K. L. Gross. 2006. The dynamic nature of ecological systems: Multiple states and restoration trajectories. Pp. 190–209 in D. A. Falk, M. A. Palmer, and J. B. Zedler, eds. *Foundations of restoration ecology*. Island Press, Washington, DC.

Suding, K. N., K. L. Gross, and G. R. Houseman. 2004. Alternative states and positive feedbacks in restoration ecology. *Trends in Ecology and Evolution* 19:46–53.

Temperton, V. M., and R. J. Hobbs. 2004. The search for ecological assembly rules and its relevance to restoration ecology. Pp. 34–54 in V. M. Temperton,

R. J. Hobbs, T. J. Nuttle, and S. Halle, eds. *Assembly rules and restoration ecology: Bridging the gap between theory and practice*. Island Press, Washington, DC.

Thompson, J. N. 1998. Rapid evolution as an ecological process. *Trends in Ecology and Evolution* 13:329–332.

Walker, B., C. S. Holling, S. R. Carpenter, and A. Kinzig. 2004. Resilience, adaptability and transformability in social-ecological systems. *Ecology and Society* 9:5.

Wanless, R. M., A. Angel, R. J. Cuthbert, G. M. Hilton, and P. G. Ryan. 2007. Can predation by invasive mice drive seabird extinctions? *Biology Letters* 3:241–244.

Wilson, K. A., E. C. Underwood, S. A. Morrison, K. R. Klausmeyer, W. W. Murdoch, B. Reyers, G. Wardell-Johnson, et al. 2007. Conserving biodiversity efficiently: What to do, where, and when. *PLoS Biology* 5(9):e223. doi:10.1371/journal.pbio.0050223:e223.

Wright, J. P., and C. G. Jones. 2006. The concept of organisms as ecosystem engineers ten years on: Progress, limitations, and challenges. *BioScience* 56:203–209.

Yachi, S., and M. Loreau. 1999. Biodiversity and ecosystem productivity in a fluctuating environment: The insurance hypothesis. *Proceedings of the National Academy of Science* 96:1463–1468.

Zavaleta, E. S., R. J. Hobbs, and H. A. Mooney. 2001. Viewing invasive species removal in a whole-ecosystem context. *Trends in Ecology and Evolution* 16:454–459.

Chapter 4

Shifting Environmental Foundations: The Unprecedented and Unpredictable Future

NATHAN L. STEPHENSON, CONSTANCE I. MILLAR, AND DAVID N. COLE

Prediction is very difficult, especially about the future.

—*Niels Bohr*

As described in Chapter 2, protected area managers have been directed, through statutes and agency policy, to preserve natural conditions in parks and wilderness. Although preserving naturalness has always been a challenge for managers, there has never been much question about whether this is the right thing to do. But given what is known now about the pace and magnitude of ongoing global changes, the appropriateness of naturalness as a management goal must be reexamined. A host of anthropogenic environmental stressors are reshaping ecosystems, including those protected in parks and wilderness. Pollution is now ubiquitous worldwide, and invasive species are common in most landscapes. Habitats have become highly fragmented, and climatic changes are dramatically altering the abiotic conditions in which biota live. Given these changes, some attempts to restore and maintain naturalness may at best be ineffective; at worst, they could waste precious resources and even contribute to loss of some of the values that managers are trying to protect.

It was once assumed that natural conditions could be maintained largely by protecting a park from development and inappropriate uses. Now, ever-expanding human impacts suggest that human intervention in ecosystems may be essential to protect critical values in parks and wilderness areas. Ecological restoration has moved toward the forefront of stewardship policy and practice. As a case in point, National Park Service (2006: 37) management policies call for the restoration of naturally functioning ecosystems and, if this is not possible, for the restoration and maintenance of "the closest approximation of the natural condition." Although restoration may enhance some of the values of naturalness, such as maintenance of native biodiversity, widespread intervention erodes other meanings and values of naturalness—natural areas as places where humans do not willfully manipulate ecosystems, where nature remains self-willed and autonomous.

Where intervention and restoration are needed and feasible, protected area managers must develop realistic objectives and devise effective strategies for achieving them. In part because of the centrality of historical fidelity to notions of naturalness, past ecosystem conditions have commonly been adopted as targets for the future. With recognition of the inherent dynamism of ecosystems (Chapter 3), reference targets for restoration are often prescribed as a range of past conditions, often called natural or historical range of variability (Landres et al. 1999), rather than conditions at a single point in time. But given the rapid pace of directional anthropogenic changes, such targets, even those expressed as a range, may be neither achievable nor desirable.

Although the range of past ecosystem conditions remains a valuable source of information about the forces that shape ecosystems (Swetnam et al. 1999), it no longer automatically serves as a sensible target for restoration and maintenance of ecosystems. Our world has entered an era in which keystone environmental drivers—those that define the possible range of characteristics of a protected area—simply have no analog in the past, no matter how distantly we look. Attempts to restore and maintain a semblance of past conditions therefore may be akin to forcing square pegs into round holes. Furthermore, at the spatial and temporal scales relevant to protected area management, the ability to predict future ecosystem conditions and outcomes of management actions is, at best, qualitative. Surprises are inevitable and are likely to be the rule rather than the exception.

In this chapter, we explore the implications of rapid global changes for protected area stewardship. We describe the major classes of anthropogenic drivers of changes in protected areas, including habitat fragmentation, loss of top predators, pollution, invasive species, altered disturbance regimes, and climatic change, concluding that resultant ecosystem changes are likely to be dramatic, ubiquitous, directional, and unprecedented. And even with management intervention aimed at responding to these stressors, protected areas inevitably will experience substantial effects of accelerating anthropogenic global changes. We describe the challenges of predicting the future, including how ecosystems are likely to respond to management actions. We identify some likely expectations for the future but conclude that uncertainty will be high. Consequently, many of the traditional approaches to protected area management that depend on natural conditions as benchmarks for restoration may no longer be tenable. We briefly point out some promising new goals and management strategies, topics that are covered in more detail in subsequent chapters.

Anthropogenic Change and the Unprecedented Future

To protect values such as native biodiversity and critical ecosystem functions, protected area managers need clear information on the nature and magnitude of anthropogenic influences on park and wilderness ecosystems. When change is deemed unacceptable and critical values are threatened, intervention will generally be needed. When interventions are taken, managers must identify desired outcomes and prescribe specific management actions likely to be effective in meeting those targets. Understanding change is fundamental to all these steps, each of which becomes increasingly difficult as anthropogenic changes increase and future conditions become less and less similar to those of the past. A handful of particularly important drivers of change have profound effects on park and wilderness ecosystems.

Habitat Fragmentation and the Loss of Top Predators

When Yellowstone National Park was first designated, it was generally assumed naturalness could be achieved by leaving the park alone. But it was soon discovered that even a large park such as Yellowstone was too small to remain natural without human intervention, particularly once top predators such as the wolf were eliminated. Loss of keystone species and processes has cascading effects that ultimately can be manifested in loss of biodiversity (Wagner 2006). And if this is a problem even in a large park such as Yellowstone, it is likely to be an even more severe problem in smaller protected areas. Increasingly, parks and wildernesses are isolated islands: relatively undisturbed biotic communities embedded in a matrix of land that has been substantially altered by humans (Hansen and DeFries 2007).

In many protected areas, top predators have been eliminated or are present in reduced numbers, cut off from other populations by adjacent developed lands. A common result is hyperabundant ungulate populations that have cascading effects throughout an ecosystem. For example, Ripple and Beschta (2008) have concluded that declining populations of black oak (*Quercus kelloggii*), an emblematic species in Yosemite Valley, may ultimately result from cougars (*Puma concolor*) now avoiding the valley, where people congregate. With predation reduced, populations of mule deer (*Odocoileus hemionus*), which browse on the oak seedlings, have expanded. Morell (2008) notes that with few small oaks to replace elders, other vegetation and animal species may be affected, potentially resulting in a decline in overall biodiversity. This decline in diversity may occur despite the natural appearance of the valley and its protection for more than a century.

Land use changes around a protected area, such as residential development, conversion to agriculture, or timber harvests, can have substantial effects on the reserve itself, such as through changes in ecological flows into and out of the reserve, loss of habitat crucial to mobile organisms, increasing exposure to invasive species along reserve edges, and changes in effective reserve size (Hansen and DeFries 2007). An example is Devil's Postpile National Monument, a small park unit in California that is surrounded by lands administered for diverse purposes by the U.S. Forest Service. Recreational uses in the multiple-use lands to the east of the monument contribute to a flow of invasive species into the adjacent portions of the monument; in contrast, a similar flow across the western boundary of the monument, which adjoins wilderness, has not been observed.

For species with poor ability to disperse across human-altered landscapes, such as species that depend on old-growth forest but that are unable to disperse across agricultural lands, habitat fragmentation may reduce genetic exchange between populations, possibly reducing adaptive potential to other novel stressors. Additionally, in the face of rapid climatic change, protected areas may become unsuitable for some of the species they protect. Habitat fragmentation by land use changes may preclude those species from migrating to new regions more suitable to them.

The Spread of Invasive Species

Another important driver of change in protected areas is invasion by nonnative species. Nonnative invasive species can substantially alter the structure, composition, and function of ecosystems. For example, chestnut blight (*Cryphonectria parasitica*) and the gypsy moth (*Lymantria dispar*) have had devastating effects in forests of eastern North America (Lovett et al. 2006). Often, the invasive species having the greatest effects on ecosystems are those that alter ecosystem processes and disturbance regimes. For example, cheatgrass (*Bromus tectorum*) has invaded the understory of many pine forests and shrublands in the western United States, increasing the amount and continuity of fine fuels and thus the frequency of fires. Increased fire frequency, in turn, affects ecosystem composition and structure and other ecosystem processes, changes that often increase vulnerability to further invasion.

Although nonnative species traditionally have been considered those that are recently arrived from anywhere outside a protected area's boundary (or even native species, such as trout, transplanted to new habitats within the boundary), rapid climatic changes will probably force a reassessment of this definition. Habitats within protected areas will probably become unsuitable for some current native species but may become suitable for species that historically may never have occurred in the protected area but that are native to the surrounding region. Such "displaced natives" may no longer automatically be treated as nonnative invasive species.

Invasive species provide important lessons regarding decisions about interventions and the relevance of naturalness to such decisions. First, most protected area managers agree that it is simply infeasible to eliminate all invasive species. For example, nonnative annual grasses have been abundant in many low-elevation park ecosystems in California since the 1850s. They are considered naturalized and are often perceived to be "natural." Interventions usually are focused on select invasive species. For example, managers intervene to fight a worrisome plant such as spotted knapweed (*Centurea maculosa*) rather than a more benign invasive because knapweed has a more deleterious effect on ecosystems (see Chapter 10 for more on invasives).

Altered Disturbance Regimes

Humans have substantially altered disturbance regimes over large parts of the earth's surface, including parks and wildernesses. For example, humans have altered fire regimes for millennia, particularly during the twentieth century. In western North America, fire exclusion after Euroamerican settlement resulted in unprecedented fire-free periods in some forest types. Lack of fire has modified forest structure and composition and increased the likelihood of wildfires sweeping through forests with a severity that was rarely encountered in pre-Euroamerican times. Fire exclusion has had cascading effects on biodiversity, biogeochemical cycles, and wildlife (Keane et al. 2002). This situation has been aggravated by climatic warming, which is implicated in longer fire seasons and increases in the area burned in large, uncontrolled wildfires (Westerling et al. 2006).

Many aquatic and riparian organisms are adapted to periodic floods, which can greatly affect habitat structure and mobilize nutrients and sediments. Dams within or upstream of some protected areas have profoundly reduced seasonal flooding, altering ecosystems and in some cases contributing to extinctions. In the face of rapid climatic changes, even unregulated rivers and streams are likely to experience altered flood regimes.

Air and Water Pollution

One of the most pervasive anthropogenic global changes, affecting both aquatic and terrestrial ecosystems, is rising atmospheric carbon dioxide (CO_2) concentrations. Although CO_2 has always been a normal part of Earth's atmosphere, human activities such as fossil fuel combustion and deforestation have led to the highest atmospheric CO_2 concentrations of any period in at least the last 650,000 years (Intergovernmental Panel on Climate Change [IPCC] 2007; Figure 4.1). Elevated CO_2 concentrations have already resulted in ocean acidification, with potentially profound consequences for marine ecosystems (Orr et al. 2005). Furthermore, rising

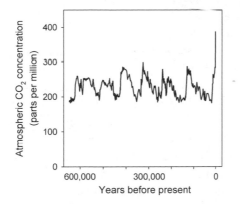

FIGURE 4.1. The nature of directional anthropogenic global changes illustrated by changes in atmospheric CO_2 concentrations through time. For at least the last 650,000 years, CO_2 concentrations have varied largely in concert with glacial advances and retreats, never falling below 180 ppm or exceeding 300 ppm. Mostly within the last 100 years, human activities such as fossil fuel combustion and deforestation have driven CO_2 concentrations well beyond 380 ppm, greatly exceeding any concentrations of at least the preceding 650,000 years. (Data from National Climatic Data Center and Oak Ridge National Laboratory)

CO_2 concentrations affect plant growth and competition, and therefore community structure and composition.

Compared with increases in atmospheric CO_2, many other forms of pollution are distributed more heterogeneously across Earth but can have regionally pervasive effects. Humans now release more biologically active nitrogen than is released by all natural processes combined, with cascading effects on ecosystems and feedbacks to climatic changes (Fenn et al. 2003; Galloway et al. 2008). Increased ground-level ozone concentrations contribute to the death or reduced growth of some plant species, thereby shifting species composition (Ashmore 2005), and acidic deposition further alters both terrestrial and aquatic ecosystems (Driscoll et al. 2001). Other widespread forms of pollution include pesticides, some of which act as endocrine disruptors.

As with invasive species, protected area managers often must accept that, at least for the foreseeable future, many of the adverse effects of pollution are inevitable. Managers can work with others to reduce pollution at its sources, but effective tools for doing so are limited. Within their protected areas, managers must identify which effects of pollution are both critically important and feasible to address. As with invasive species, priorities for action must be based not on how unnatural a pollutant is but on how critical its effect is to desired ecosystem characteristics and the likely efficacy of proposed management actions.

Global Climatic Change

As noted in Chapter 2, recognition of the magnitude of global climatic change has had the most profound effect on how we think about protected area stewardship and the relevance of naturalness to stewardship decisions. Earth may be warmer now than at any time in the last millennium, and perhaps much longer. If expected future climatic changes come to pass, by the end of this century large portions of the earth's surface will host climates that have no current analog anywhere on Earth (Williams and Jackson 2007). If we consider specific combinations of climate, soils, and topography as defining a particular habitat type, ongoing climatic changes may result in even greater proportions of the earth's surface being occupied by novel habitats than would be expected if climatic changes alone were considered (Saxon et al. 2005). Conversely, climatic changes are also expected to result in the complete disappearance of some contemporary climates and habitat types, both regionally and globally (Saxon et al. 2005; Williams and Jackson 2007).

Anthropogenic climatic change will drive park and wilderness eco-systems further from the natural condition mandated by policy, in which human influence is minimal and there is a substantial degree of historical fidelity. Currently, managers are often directed to respond by restoring as "natural" a condition as possible. But what do they use as a target or refer-ence for their interventions if the future climate of a particular protected area is projected to have no contemporary analog? Can they expect to find a reference analog in Earth's past? Perhaps, but that past may be quite dis-tant and its ecological setting quite different. For example, by the end of this century average global temperatures may have reached levels not seen since the last interglacial period or longer (more than 120,000 years ago), a period when now-extinct megafauna played significant roles as ecosystem architects. By century's end, global temperatures might even exceed any achieved in the last several million years, evolutionary time scales that have seen pervasive changes in Earth's biota. Even if it were possible to charac-terize the structure, composition, and processes of ancient ecosystems with adequate precision to use them as references, their biota differed from those on Earth today.

Although the extent of no-analog habitats (combinations of climate, soils, and topography) is likely to increase in coming decades, the altered future habitats of some protected areas might sometimes be analogous to habitats found elsewhere on the landscape within the past few centuries, particularly in environmentally complex mountainous regions (Saxon et al. 2005). Would such protected areas be sensible candidates for contin-ued management for "natural" conditions, using analogs from the past as targets? We think not. A site's biotic potential in the future will be deter-mined by more than just climate, soils, and topography. Specifically, cli-matic changes will interact with the other novel, pervasive agents of change (habitat fragmentation, loss of top predators, invasive species, altered dis-turbance regimes, and pollution) to such an extent that all habitats may be no-analog habitats.

Unprecedented Environmental Change Challenges Naturalness

The unprecedented future that will result from the convergence of rapid climatic changes with an additional suite of novel, pervasive environmental stressors such as those described above demands that managers and poli-cymakers move beyond existing concepts of naturalness. Anthropogenic change is both ubiquitous and directional, and restoration of key aspects of naturalness (such as historical fidelity) is likely to be both unattainable and undesirable. However, directional change does not mean that change is either linear or predictable. In fact, as will be discussed in the next section, change may be nonlinear and increasingly unpredictable. We have entered an era in which environmental influences on ecosystems have no precedence in the history of Earth, no matter how far into the past we look.

Our entry into a no-analog future means that attempts to restore naturalness, using past conditions as targets, usually will demand greater and greater inputs of energy. Protected area managers may thus be committed to engage in Sisyphean efforts that ultimately are likely to fail (Hilderbrand et al. 2005). Even more significantly, the no-analog future means that management interventions could result in "restored" ecosystems that are inherently unstable to novel conditions, making them more susceptible to sudden, undesirable state shifts (Harris et al. 2006). For example, maintenance of a "naturally" dense forest in the face of a drying climate could result in the sudden loss of the forest to insects, pathogens, or an unusually severe wildfire, followed by soil erosion and a consequent reduction of biological potential. Thus, in the face of a suite of novel environmental conditions, restoration of ecosystems to resemble those of the past provides no guarantee of their sustainability into the future and in fact might lead to the catastrophic loss of some of the very ecosystem elements intended for preservation.

The Unpredictable Future

To decide whether and how to intervene in ecosystems, protected area managers normally need a reasonably clear idea of what future ecosystems would be like if they did not intervene. Management practices usually involve defining a more desirable future condition and implementing management actions designed to push or guide ecosystems toward that condition. Managers need confidence in the likely outcomes of their interventions. This traditional and inherently logical approach requires a high degree of predictive ability, and predictions must be developed at appropriate spatial and temporal scales, often localized and near-term. Unfortunately, at the scales, accuracy, and precision most useful to protected area management, the future not only promises to be unprecedented, it promises to be unpredictable. To illustrate this, consider the uncertainties involved in predicting climatic changes, how ecosystems are likely to respond to climatic changes, and the likely efficacy of actions that might be taken to counter adverse effects of climatic changes. Comparable uncertainties surround the nature and magnitude of future changes in the other ecosystem stressors we have discussed in this chapter and the interactions between these stressors.

The Challenge of Predicting Climatic Changes

Prediction of the rate and magnitude of future climatic changes requires knowledge of climate sensitivity, the average global warming expected from a doubling of atmospheric CO_2 concentrations. Recent estimates of climate sensitivity range broadly between 2.1° and 4.4° C, although lower and much higher values are possible. Additional research is unlikely to improve this range of estimates because the large degree of uncertainty is a general consequence of the nature of the climate system itself. Thus, our ability to predict the effects of increasing greenhouse gas concentrations even on global-scale climate is inherently uncertain.

This uncertainty is compounded by uncertainty about the magnitude of future greenhouse gas emissions. Emissions will be affected by the interactions of complex phenomena such as long-term demographic trends, economic development, land use change, technological change, geopolitics, and feedbacks between climatic changes, ecosystems, and societies. Recent international climatic change assessments have relied on greenhouse gas emission scenarios spanning a broad range of possible futures. Even so, actual emission reductions from technological advances have already lagged behind projected reductions that were assumed in even the most pessimistic of these scenarios (Pielke et al. 2008), underscoring the inherent uncertainty in predicting future emissions.

Uncertainty regarding both greenhouse gas emissions and climatic response to those emissions means that the rate and magnitude of future global climatic changes likewise remain uncertain. For example, recent estimates of increases in mean global temperature by the end of the century mostly range from 1.1° to 6.4° C, with even higher or lower values possible (IPCC 2007). Importantly, these are globally averaged predictions. The accuracy of climatic change predictions decreases as the scale of analysis is narrowed from global to regional to local. Local predictions, the most inaccurate ones, are what the managers of protected areas really need. Even for well-studied regions such as northern California, recent model projections do not agree as to whether future climates will be warmer and wetter or warmer and drier—alternative futures that have profoundly differing implications for protected areas and their management. At the spatial scales relevant to park managers, even higher levels of uncertainty arise from poorly understood microclimatic complexity, such as cold air pooling and hillslope effects.

The Challenge of Predicting Ecosystem Response to Climatic Change

Climate forecasts are relevant because those forecasts then drive models addressing the questions of greatest interest to managers: How will the plants, animals, and ecosystems they steward be affected by future climatic changes? Yet predictions of ecosystem responses to environmental changes are notoriously unreliable. Their accuracy would be low even if the rate and magnitude of future environmental changes were precisely known.

Models are useful for organizing thinking, giving a qualitative feel for a range of possible rates and magnitudes of future ecosystem changes, and providing grist for scenario planning (Chapter 13). However, their outputs cannot be used as predictions. Because the extraordinary complexity of ecosystems is not fully understood and cannot be adequately incorporated in models, models make numerous simplifying assumptions. For example, some common assumptions are that the effects of species interactions are negligible, that evolutionary responses to rapid environmental changes are negligible, and that contemporary correlations between environment and ecosystem properties imply simple cause-and-effect relationships. However, most such assumptions are questionable or are simply false (Dormann 2007; Suttle et al. 2007). Models based on different sets of simplifying assumptions yield widely divergent forecasts of the biological effects of future climatic changes. For example, projected changes in distribution of the South African shrub *Leucospermum hypophyllocarpodendron* between now and 2030 ranged from a 92 percent reduction to a 322 percent gain over the plant's current distribution, depending on which of several competing models was used (Pearson et al. 2006). Although forecasts might improve by considering the combined outputs of several models, significant uncertainties and surprises are inevitable (Williams and Jackson 2007; Doak et al. 2008).

Further hampering prediction of future ecosystem conditions, environmental stressors often interact in unexpected ways. For example, atmospheric nitrogen deposition can facilitate invasion by nonnative plant species, which in turn can alter fire regimes, ultimately leading to vegetation type conversions (Brooks et al. 2004). Such interactions are notoriously difficult to anticipate (Doak et al. 2008). Other surprises will occur as additional stressors emerge, a recent example being the rapid, global spread of a chytrid fungus (*Batrachochytrium dendrobatidis*) throughout frog populations. Effects of such diseases can trigger cascading effects throughout ecosystems.

A critically important class of surprises involves threshold events, in which gradual environmental changes trigger sudden, dramatic, and sometimes irreversible changes in ecosystem state (Scheffer and Carpenter 2003). For example, in some parts of western North America gradual climatic warming has contributed to sudden and extensive outbreaks of bark beetles (*Dendroctonus* and *Ips*), killing millions of hectares of forest within a few years (Raffa et al. 2008). Ongoing warming may trigger further outbreaks in regions formerly immune to them. Although critically important, thresholds remain difficult to identify (Scheffer and Carpenter 2003).

Some Broad Expectations for Future Conditions

Although the future is impossible to predict precisely, particularly at the local scales most important to managers, it is still possible to identify broad, qualitative expectations at regional scales that might be helpful to protected area managers. For example, even though the precise pace and magnitude of global or regional warming cannot be predicted, one can predict with high certainty that in coming decades most regions of the earth will get warmer (Jackson and Overpeck 2000). Thus, it is reasonable to expect that at the regional if not local scale, the amount of annual precipitation falling as snow will continue to decline and that snowpack itself will decline and, in some places, vanish. As summers become longer, the average regional fire season will lengthen, and area burned will probably increase (Westerling et al. 2006). Weather extremes, such as heat waves, droughts, and floods, are likely to become more common (IPCC 2007). Sea level will continue to rise, inundating some coastal ecosystems, and oceans will continue to warm and acidify (IPCC 2007). Land use changes are likely to continue driving habitat fragmentation, various sorts of pollution are likely to continue or increase, and novel invasive species will continue invading most ecosystems.

Paleoecological studies have clearly demonstrated that species behave individualistically in response to climatic changes (Jackson and Overpeck 2000). Thus, in response to climatic changes of the magnitude projected by the end of this century, most contemporary biotic communities (particular combinations of species that currently live together) are likely to have at least partly dissociated, and their component species will have reassembled in combinations that have no contemporary analogs. As alluded to earlier, either with or without human assistance, many species will migrate, with ranges expanding in some areas and contracting in others (Parmesan 2006; Thomas et al. 2006). Other species will almost certainly be driven to population extirpation or species extinction (Thomas et al. 2006). Many protected areas will no longer provide suitable habitat for many of their current species; conversely, "displaced native" species are likely to migrate into reserves. Rapid environmental changes will also drive evolutionary changes, altering biodiversity at the level of the genome (Parmesan 2006).

Unpredictability Makes Desired Future Conditions Problematic

In the next several decades, ongoing global environmental changes are almost certain to drive profound and unprecedented changes in ecosystems. However, the precise nature of those changes will not be clear until they happen. Despite the scientific community's ability to make some broad, qualitative generalizations about probable future conditions, it is impossible, especially at the spatial and temporal scales useful to protected area managers, to accurately predict either environmental changes or consequent ecosystem responses (Pilkey and Pilkey-Jarvis 2007). Given this uncertainty, narrowly defined desired future ecosystem conditions, particularly if they are historical conditions poorly aligned with the unprecedented future, will seldom provide useful targets for management intervention. Managers will need approaches to planning that are more suitable to a high level of uncertainty, approaches that allow for a much wider array of possible desired outcomes, and they will need ways to become even more adept at adapting to rapidly changing conditions.

Protected Area Stewardship in a Changing World

The nature and magnitude of anthropogenic global changes challenges managers to move beyond traditional concepts of naturalness to find guidance for dealing with the nuanced dilemmas of contemporary stewardship. The degree to which protected area ecosystems are affected by humans will almost certainly continue to increase, and it is probable that the pervasiveness of management interventions will follow suit. Although there will always be places where lack of funds precludes intervention and places where managers refrain from intervening despite human effects on ecosystems, many protected areas are likely to experience at least some intentional human manipulation. The numerous anthropogenic drivers of change and the profound nature of change make it clear that substantial human influence is inevitable, even in our most valued parks and wilderness areas.

The future dominance of no-analog conditions, the inherent unpredictability of the future, and the virtual certainty of surprises all conspire to force protected area managers to adopt fundamentally new goals and management strategies. As we have noted, management actions aimed at restoring or maintaining ecosystems with high historical fidelity (using natural analogs from the past) or mimicking conditions that would exist in the absence of human influences will require continuously increasing inputs of energy and ultimately are likely to fail. Failures could be catastrophic. Attempts to resist rapid environmental changes could create inherently unstable conditions, leading to the sudden loss of some of the very species and ecosystem functions managers had hoped to sustain. Instead, protected area managers will probably need to redefine their goals. For example, they might seek to maintain regional native biodiversity and critical ecosystem functions, even if biotic community structure and composition no longer resemble what existed in the past. To reach these ends, new management approaches are needed. Interventions might emphasize facilitating, rather than resisting, certain ecosystem changes. A key to facilitating ecosystem transitions will be maintenance of ecosystem resilience, the ability to sustain environmental shocks and stresses without undergoing an undesirable and irreversible change in conditions. Resilience and other management emphases are addressed in Chapters 6 through 9.

Managers will need to act in the face of uncertainty. To do so, they may need to abandon traditional approaches to long-term planning that are based on the assumption that the future is known, or at least knowable. They are not likely to be able to use the outputs of computer models to determine a particular, narrowly defined future ecosystem trajectory to facilitate. Rather than attempt to define a specific set of desired future conditions, it may sometimes be more productive to define a broad set of undesired future conditions—conditions to be avoided. For example, undesired future conditions might include loss of regional native biodiversity or critical ecosystem functions. Outcomes that do not fall within the undesired future conditions may be deemed acceptable. In the face of high uncertainty, managers might engage in scenario planning (described in Chapter 13), the use of internally consistent visions of a range of possible futures to explore potential future consequences of different decisions. For example, scenario planning might suggest that a particular set of management actions could lead to ecosystems resilient to a wide variety of potential stresses, including both warmer and wetter and warmer and drier futures.

The anthropogenic threats to protected areas discussed in this chapter present profound challenges to protected area stewardship. These challenges are exacerbated by the unprecedented future that protected areas face and the limited ability of science to predict the future and the likely outcomes of management interventions. However, as challenging as our assessments may initially sound, we firmly believe that scientists, managers, and policymakers can work together to define new ways to protect key values of parks and wilderness areas: new goals, new institutions, new planning processes, and new management approaches, innovations discussed more fully in the chapters that follow.

BOX 4.1. GLOBAL ANTHROPOGENIC CHANGE AND NATURALNESS

- Anthropogenic forces—most notably habitat fragmentation and loss of top predators, invasive species, altered disturbance regimes, pollution, and climatic change—are having profound effects on protected areas.
- These profound, ubiquitous, and directional changes will lead to an unprecedented future for which there is no analog, now or in the past.
- Although useful sources of information for many purposes, past ecosystem conditions (historical or natural range of variability) often will make poor targets for management interventions.
- Despite management interventions, protected area ecosystems inevitably will suffer the effects of anthropogenic global changes and will lose much of their historical fidelity.
- Consequently, simple and traditional concepts of naturalness will prove inadequate to provide guidance regarding stewardship decisions about where and how to intervene in ecosystems.
- In addition to being unprecedented, the future will also be largely unpredictable at the scales useful to managers and full of uncertainty and surprises.
- Because the outcomes of management interventions will be uncertain, protected area managers will have to be adaptable, investing more in experimenting, monitoring, and learning and in more flexible approaches to planning.

REFERENCES

Ashmore, M. R. 2005. Assessing the future global impacts of ozone on vegetation. *Plant, Cell and Environment* 28:949–964.

Brooks, M. L., C. M. D'Antonio, D. M. Richardson, J. B. Grace, J. E. Keeley, J. M. DiTomaso, R. J. Hobbs, M. Pellant, and D. Pyke. 2004. Effects of invasive alien plants on fire regimes. *BioScience* 54:677–688.

Doak, D. F., J. A. Estes, B. S. Halpern, U. Jacob, D. R. Lindberg, J. Lovvorn, D. H. Monson, et al. 2008. Understanding and predicting ecological dynamics: Are major surprises inevitable? *Ecology* 89:952–961.

Dormann, C. F. 2007. Promising the future? Global change projections of species distributions. *Basic and Applied Ecology* 8:387–397.

Driscoll, C. T., G. B. Lawrence, A. J. Bulger, T. J. Butler, C. S. Cronan, C. Eagar, K. F. Lambert, G. E. Likens, J. L. Stoddard, and K. C. Weathers. 2001. Acidic deposition in the northeastern United States: Sources and inputs, ecosystems effects, and management strategies. *BioScience* 51:180–198.

Fenn, M. E., J. S. Baron, E. B. Allen, H. M. Rueth, K. R. Nydick, L. Geiser, W. D. Bowman, et al. 2003. Ecological effects of nitrogen deposition in the western United States. *BioScience* 53:404–420.

Galloway, J. N., A. R. Townsend, J. W. Erisman, M. Bekunda, Z. Cai, J. R. Freney, L. A. Martinelli, S. P. Seitzinger, and M. A. Sutton. 2008. Transformation of the nitrogen cycle: Recent trends, questions, and potential solutions. *Science* 320:889–892.

Hansen, A. J., and R. DeFries. 2007. Ecological mechanisms linking protected areas to surrounding lands. *Ecological Applications* 17:974–988.

Harris, J. A., R. J. Hobbs, E. Higgs, and J. Aronson. 2006. Ecological restoration and global climate change. *Restoration Ecology* 14:170–176.

Hilderbrand, R. H., A. C. Watts, and A. M. Randle. 2005. The myths of restoration ecology. *Ecology and Society* 10(1):19. Retrieved September 22, 2009 from www.ecologyandsociety.org/vol10/iss1/art19/.

Intergovernmental Panel on Climate Change. 2007. *Climate change 2007: Synthesis report.* Retrieved September 22, 2009 from www.ipcc.ch/pdf/assessment-report/ar4/syr/ar4_syr.pdf.

Jackson, S. T., and J. T. Overpeck. 2000. Responses of plant populations and communities to environmental changes of the late Quaternary. *Paleobiology* 26(Suppl.):194–220.

Keane, R. E., K. C. Ryan, T. T. Veblen, C. D. Allen, J. A. Logan, and B. Hawkes. 2002. The cascading effects of fire exclusion in Rocky Mountain ecosystems. Pp. 133–152 in J. S. Baron, ed. *Rocky Mountain futures: An ecological perspective.* Island Press, Washington, DC.

Landres, P. B., P. Morgan, and F. J. Swanson. 1999. Overview of the use of natural variability concepts in managing ecological systems. *Ecological Applications* 9:1179–1188.

Lovett, G. M., C. D. Canham, M. A. Arthur, K. C. Weathers, and R. D. Fitzhugh. 2006. Forest ecosystem responses to exotic pests and pathogens in eastern North America. *BioScience* 56:395–405.

Morell, V. 2008. Yosemite: Protected but not preserved. *Science* 320:597.

National Park Service. 2006. *Management policies 2006.* www.nps.gov/policy/MP2006.pdf.

Orr, J. C., V. J. Fabry, O. Aumont, L. Bopp, S. C. Doney, R. A. Feely, A.

Gnanadesikan, et al. 2005. Anthropogenic ocean acidification over the twenty-first century and its impact on calcifying organisms. *Nature* 437:681–686.

Parmesan, C. 2006. Ecological and evolutionary responses to recent climate change. *Annual Review of Ecology, Evolution, and Systematics* 37:637–669.

Pearson, R. G., W. Thuiller, M. B. Araújo, E. Martinez-Meyer, L. Brotons, C. McClean, L. Miles, P. Segurado, T. P. Dawson, and D. C. Lees. 2006. Model-based uncertainty in species range prediction. *Journal of Biogeography* 33:1704–1711.

Pielke, R. Jr., T. Wigley, and C. Green. 2008. Dangerous assumptions. *Nature* 452:531–532.

Pilkey, O. H., and L. Pilkey-Jarvis. 2007. *Useless arithmetic: Why environmental scientists can't predict the future*. Columbia University Press, New York.

Raffa, K. F., B. H. Aukema, B. J. Bentz, A. L. Carroll, J. A. Hicke, M. G. Turner, and W. H. Romme. 2008. Cross-scale drivers of natural disturbances prone to anthropogenic amplification: Dynamics of bark beetle eruptions. *BioScience* 58:501–517.

Ripple, W. J., and R. L. Beschta. 2008. Trophic cascades involving cougar, mule deer, and black oaks in Yosemite National Park. *Biological Conservation* 141:1249–1256.

Saxon, E., B. Baker, W. W. Hargrove, F. M. Hoffman, and C. Zganjar. 2005. Mapping environments at risk under different global climate change scenarios. *Ecology Letters* 8:53–60.

Scheffer, M., and S. R. Carpenter. 2003. Catastrophic regime shifts in ecosystems: Linking theory to observation. *Trends in Ecology and Evolution* 18:648–656.

Suttle, K. B., M. A. Thomsen, and M. E. Power. 2007. Species interactions reverse grassland response to changing climate. *Science* 315:640–642.

Swetnam, T. W., C. D. Allen, and J. L. Betancourt. 1999. Applied historical ecology: Using the past to manage for the future. *Ecological Applications* 9:1189–1206.

Thomas, C. D., A. M. A. Franco, and J. K. Hill. 2006. Range retractions and extinction in the face of climate warming. *Trends in Ecology and Evolution* 21:415–416.

Wagner, F. H. 2006. *Yellowstone's destabilized ecosystem: Elk effects, science, and policy conflict*. Oxford University Press, New York.

Westerling, A. L., H. G. Hidalgo, D. R. Cayan, and T. W. Swetnam. 2006. Warming and earlier spring increase western U.S. forest wildfire activity. *Science* 313:940–943.

Williams, J. W., and S. T. Jackson. 2007. Novel climates, no-analog communities, and ecological surprises. *Frontiers in Ecology and the Environment* 5:475–482.

Chapter 5

Changing Policies and Practices: The Challenge of Managing for Naturalness

LAURIE YUNG, DAVID N. COLE, DAVID M. GRABER,
DAVID J. PARSONS, AND KATHY A. TONNESSEN

It can be safely said that when it comes to actual work on the ground,
the objects of conservation are never axiomatic or obvious but always
complex and usually conflicting.

—*Aldo Leopold*

Protected area policy and practice have changed dramatically over the past
century, in response to shifting societal values, conservation politics, and scientific understanding, and ever-increasing human environmental impacts.
Public enjoyment and scenic beauty were once the highest priority in U.S.
national parks. At the start of the twentieth century, only the "desirable"
native species were protected, while others were exterminated; "undesirable" ecosystem elements, such as fire, were controlled wherever possible.
But by the latter half of the twentieth century, parks and wilderness began
to embrace all native species and ecosystem processes, and protected areas
became increasingly viewed as critical cornerstones of biodiversity conservation. At the same time, conservation advocates argued that active management should be kept to a minimum, to allow nature to take its course
free from human intervention.

Chapter 2 in this book examined the multiple meanings of naturalness
and how these lead to conflicting goals for protected areas. Chapters 3 and

4 explored how new scientific insights and an expanding global human imprint suggest that protected area stewardship is increasingly complex and challenging and that concepts such as naturalness are inadequate for dealing with the nuances of managing ecosystems in parks and wilderness. In this chapter we explore how protected area policy and management practice have struggled to implement the mandate of naturalness. We argue that management for naturalness has proved particularly thorny, in large part because of vague and ambiguous policy. We focus here on national parks, but the challenges described apply to many protected areas in the United States and internationally.

From Scenery and Spectacle to Biodiversity Conservation

National parks were initially established to protect dramatic scenery and charismatic wildlife from exploitation and modern development (Runte 1979). The National Park Service Organic Act (1916) established parks "to conserve the scenery and the natural and historic objects and the wild life therein and to provide for the enjoyment of the same in such manner and by such means as will leave them unimpaired for the enjoyment of future generations." Though a radical notion in the face of development on surrounding lands, the parks were "viewed mainly as scenic pleasuring grounds" (Sellars 1997: 27). Early park management focused on scenery and spectacle for public enjoyment and entertainment (Graber 2003). Because scenery was the primary inspiration for the establishment of early parks, as well as a powerful social symbol, aesthetics were often given priority during the early years. In some places, this "façade management" (Sellars 1997) was highly manipulative and invasive; furthermore, the management of ecological elements in parks was only minimally informed by science or conservation goals.

Throughout the first half of the twentieth century, the Park Service actively managed to enhance desirable fish and wildlife. Predators were controlled to increase deer and elk populations, bears were fed to entertain visitors, nonnative fish were introduced to enhance fishing opportunities, and ungulates were either fed during difficult winters to prevent die-offs or culled to prevent overpopulation. "Undesirable" ecosystem elements, such as predators, pathogens, fires, and floods, were controlled or eliminated where possible. Management intervention favored certain species and ecological processes, with the goal of making parks beautiful, fascinating, and pleasant, consistent with park practices that promoted tourism and encouraged development of visitor facilities. Tracts of undeveloped backcountry and preservation of scenic beauty were believed to satisfy the Organic Act mandate that parks remain "unimpaired."

By the 1930s, however, park management was expanding to include initial attempts at nature preservation (Sellars 1997). As described in Chapter 2, a National Park Service wildlife division was established, under the leadership of biologist George Wright. Although it was short-lived, Wright and others in the wildlife division argued that parks should promote native species rather than nonnative species, stop controlling native predators, and minimize overly manipulative wildlife management practices (e.g., rearing bison [*Bison bison*] or feeding elk [*Cervis canadensis*] in winter). Of particular importance, they argued that all native species—not just charismatic ones—should be protected and perpetuated.

In 1931, the National Park Service declared that all animals would find refuge in the national parks (Pritchard 2002: 50). However, the wildlife division was largely unsuccessful in its attempt to change park policy and practice. As Sellars points out, there was not yet sufficient public support for such changes.

Expanding Statutory Guidance

By the 1960s, social values were changing, and ecological concepts had gained significant traction in science and society. Public interest in environmental issues had increased, and environmental groups were becoming politically powerful. A growing concern for conservation translated into new goals for parks and wilderness. Environmental legislation from the 1960s and early 1970s expanded park purposes, sending managers in increasingly diverse directions. In 1964, the Wilderness Act established a system of wilderness areas, many of which ultimately were designated within national parks. The language of the act paralleled park policies, stating that a wilderness area should retain "its primeval character and influence." But, as discussed in Chapter 2, wilderness was also defined as a place that was "untrammeled by man," a place where humans did not intervene in ecosystem processes. The Wilderness Act codified the emerging interest in wildness, the desire to protect autonomous nature or self-willed ecosystems (Cole 2000) (see Chapter 6). The act reflected the sentiment that there should be some places where nature is left alone. A hands-off management approach was believed to be both a sign of humility and a tool for nature conservation.

A few years later, the National Environmental Policy Act (NEPA) (1969) codified a broad public interest in environmental protection. NEPA required that agencies consider the environmental impacts of federal decisions and, as part of the planning process, commit to a specific course of action for the foreseeable future. NEPA also mandated public participation and empowered citizens to challenge agency decisions. Thus, the public began to play an increasingly important role in park and wilderness planning, and a much broader set of values and interests were considered in decision making. The shift toward increased public participation in many ways challenged the notion that decisions about public lands should be made by professional experts housed within the agencies and empowered the public to participate in a planning process designed to implement broad policy guidance in particular places.

In 1973, in response to growing public concern over species extinction and loss of habitat, the Endangered Species Act (ESA) was enacted. The ESA directed the federal government (and some private landowners) to preserve habitat for and recover populations of threatened and endangered species. Although parks continued to emphasize scenic beauty and public enjoyment, both parks and wilderness now played a critical role in biodiversity conservation, providing habitat for efforts to recover threatened and endangered species. Parks had always been about conserving select elements of biodiversity (hence the focus on "wild life" in the Organic Act); increasingly, parks were seen as places that should conserve all native biodiversity, from the smallest thermophilic bacterium to the tallest redwoods.

For example, the ESA was instrumental in enabling the reintroduction of wolves (*Canis lupus*) in Yellowstone National Park and central Idaho wilderness areas in the 1990s. Science provided insights into the ecological role of predators in ecosystem structure and function, which in turn improved public understanding of predators and contributed to public support. But public support for wolf reintroduction has been far from unanimous, and wolf management continues to be controversial, especially in the rural West. Sociopolitical factors, such as livestock producers' experience of predation and hunting outfitters' concerns about declining elk populations, continue to influence how predators are managed in and adjacent to protected areas. Emerging conservation values, as expressed in new policies such as the Wilderness Act or ESA, continue to be contested and challenged, both in broad public debate and in policymaking circles.

Park Service Natural Resource Management Policy

The social, political, and scientific changes that inspired new environmental laws eventually catalyzed changes to National Park Service policy. For decades, National Park Service management practices and policies evolved in a largely ad hoc manner. In 1962, however, amid ongoing controversy over elk reductions in Yellowstone National Park, Secretary of Interior Stewart Udall commissioned wildlife biologist A. Starker Leopold to lead a panel of scientists in a formal review of national park wildlife policy. National Park Service historian Sellars (1989) argues that the Leopold report (Leopold et al. 1963) was the most influential policy statement on park management since the 1916 National Park Service Organic Act.

The Leopold report affirmed that the objects of preservation in national parks were native ecological systems in all their complexity—not nonnatives and not just certain native species but entire systems, including ecosystem processes. More specifically, the report advocated the restoration of native species and the removal of exotic plants and animals. It suggested that historic sources of disturbance, such as fire, insects, and disease, were critical ecosystem elements that should be embraced. The report also argued that, to be successful, park management must be supported by a strong base of science and research. The report said that "the maintenance of naturalness should prevail," that many park ecosystems were damaged and in need of restoration, and that restoration would require active intervention and could "not be done by passive protection alone" (Leopold et al. 1963: 34–35). The report reflected contemporary ecological thinking, recognizing the importance of disturbance processes and the dynamism of ecosystems.

But the Leopold report sent a mixed message on a number of important issues. Although it noted the dynamism of ecosystems, the report recommended that national parks represent a "vignette of primitive America" and that parks should preserve or recreate "the condition that prevailed when the area was first visited by the white man" (Leopold et al. 1963: 33). This left individual park managers to decide whether it was sufficient to simply allow for the free play of ecosystem processes or whether they should more aggressively intervene to restore structural elements or even some past state. Although it advocated active habitat manipulation, the report noted that artificiality should be minimized, that regulation should be as much as possible "by natural means." Again, individual managers were left to decide when the need for active manipulation justified increased artificiality. Indeed, the report states, "The major policy change which we would recommend to the National Park Service is that it recognize the enormous complexity of ecologic communities and the diversity of management procedures required to preserve them" (Leopold et al. 1963: 34). In essence, the report did little to resolve the divergent meanings of *naturalness* discussed in Chapter 2 of this book. It argued that park managers needed flexibility and individual discretion to address the stewardship issues unique to their park and situation, but it provided little guidance beyond general notions such as maintaining naturalness. The Leopold report assumed that a higher level of scientific sophistication in itself would guide managers to the actions needed to maintain desired conditions. But as the management struggles outlined in this chapter demonstrate, this was not the case.

In 1967, many of the Leopold report's recommendations formed the basis for a manual of policies for managing undeveloped areas in the national parks (Sellars 1997). Not surprisingly, managing for naturalness proved much more challenging than anticipated by Leopold and fellow committee members.

The Challenges of Naturalness in Practice

Two prominent issues, management of wild ungulate populations and fire, provide insights into the challenges of managing for naturalness. Each of these management dilemmas raises questions about whether and how to intervene when an ecological process has been removed from the landscape — predator–prey relationships in one case and fire in the other. Each case illustrates that social and political factors are major drivers of park practice, despite the Leopold report's assertion of the primacy of ecologically based management. Each demonstrates the ambiguity of park policy, even after the Leopold report, regarding when intervention is necessary, how to balance the dynamism of ecosystems with a desire to maintain historical fidelity (vignettes of primitive America), and whether management outcomes should emphasize state or process. In each case, resultant problems were exacerbated by incongruities between the small size of protected areas and the large spatial scale of many ecological processes (Porter and Underwood 1999), despite the fact that these issues were most hotly debated in some of the largest national parks.

Debate over the management of ungulates and fire grew in the 1980s, just as ecological understandings of protected areas were beginning to shift. As described in Chapter 3, notions of climax ecosystems, homeostasis, equilibrium, and carrying capacity were being overturned, replaced by an increased emphasis on disturbance, heterogeneity, and flux. Ecosystem dynamism meant that the "vignette" that the Leopold report described was, in fact, a moving target. Thus, early efforts to restore naturalness needed to do more than look backward to pre-Columbian conditions. In thinking about ecological processes such as fire, scientists and managers had to grapple with disturbance ecology and consider how an understanding of disturbance, such as insect outbreaks or flooding, might influence management practice (Christensen 1988). And as research documented the ways in which Native Americans had dramatically influenced certain ecosystems, it became clear that what early European settlers saw as nature free from human impact was often a highly managed system that had co-evolved with human activities (Mann 2005). Thus, managers also had to consider whether some anthropogenic forces should be considered natural.

Management of Wild Ungulate Populations

Wild animals are among the most valued national park resources. For many park visitors, glimpses of free-roaming ungulates rank among the highlights of their trip. However, ungulate populations also cause problems in parks. In some eastern parks, deer populations have increased greatly in response to reductions in predator populations and the elimination of hunting, resulting in dramatic changes to vegetation (Porter and Underwood 1999). In Yellowstone National Park, management of elk has been highly controversial. Elk populations have fluctuated dramatically since the park was established, with resultant effects on vegetation and other animal species (Huff and Varley 1999; Wagner 2006).

There have been several distinct phases in the management of the northern elk herd in Yellowstone (Huff and Varley 1999; Wagner 2006), varying in their degrees of management intervention and desired outcomes. After the removal of Native Americans from Yellowstone and the surrounding area and the creation of the park in 1872, the herd was protected from hunting and predators and provided with artificial feed during winter, which allowed elk numbers to increase from 20,000 to 35,000 animals in the early 1900s (Wagner 2006). By the 1930s, concerns about vegetation damage, attributed to an enlarged elk herd and suggestive of an overabundance of ungulates, led to a reversal of this policy. Managers intervened to compensate for the loss of natural predators by culling or removing elk to reduce numbers. From 1935 through 1967, Yellowstone elk were trapped and live-shipped to other elk ranges or shot by National Park Service hunters, leaving a northern herd population in the late 1960s of just 3,000 to 4,000 (Huff and Varley 1999; Wagner 2006).

Although the influential Leopold report concluded that shooting excess ungulates in Yellowstone was necessary, public outcry over elk reduction operations intensified in the 1960s, leading to congressional hearings on the issue (Sellars 1997). In 1967, the Park Service announced it would cease killing elk in the parks and adopt a new management approach called natural regulation or hands-off natural process management (Huff and Varley 1999). This noninterventionist approach was predicated on the hypothesis that elk populations are self-regulating (Wright 1999), that herd numbers are controlled by environment and available food.

Despite 35 years and several million dollars spent on research into the Yellowstone northern herd and the northern range, scholarly accounts both criticize and defend the natural regulation approach to ungulate management (e.g., National Academy of Sciences 2002; Singer et al. 1998; Wagner et al. 1995). In the 1990s, the conditions changed dramatically when wolves were reintroduced to Yellowstone Park. Reintroduction of wolves, it was hypothesized, might result in changes both to the elk population and to vegetation (Smith et al. 2003). Although it is too early to definitively evaluate the effects of wolf reintroduction (Wagner 2006), there is some evidence of positive ecological effects. Studies report that since wolf reintroduction, elk populations have declined somewhat, and growth of both aspen (*Populus tremuloides*) and willow (*Salix* sp.)—trees adversely affected by excessive elk browsing—has increased as fear of predation alters the spatial distribution and foraging behavior of elk, creating local refugia for tree and shrub species (Ripple et al. 2001; Beyer et al. 2007).

The ambiguity of the naturalness concept is apparent insofar as each of these approaches to elk management has been justified in the name of protecting natural conditions. When managers thought elk populations were declining, they intervened to increase herd size. Then, out of fear that natural vegetation was threatened by hyperabundant elk herds, managers intervened to reduce populations. A few decades later, the National Park Service embraced the notion of natural regulation. More recently, managers intervened to restore a missing top predator, the wolf, perhaps increasing the likelihood that ongoing intervention would not be necessary. At the same time, public outrage over shooting elk in Yellowstone may have been more important than ecological science in overturning decades of active ungulate management in national parks (Wright 1999).

Fire Management in National Parks

Since the early twentieth century, most fires in national parks have been suppressed. By the 1960s, there was increasing awareness among scientists and some park managers that suppressing fires eliminated an important disturbance process, critical to ecosystem function. In the mixed conifer forest of the Sierra Nevada parks, for example, suppression prevented re-production by giant sequoia (*Sequoiadendron giganteum*) and led to dense growth of shade-tolerant trees, such as white fir (*Abies concolor*). Fuel loads increased, transforming a fire regime characterized by frequent, relatively low-intensity surface fires into a regime of infrequent crown fires, capable of destroying ancient sequoia trees, symbols of the Sierran parks (Graber 1995). In 1968, the National Park Service, in a revision of management policies, recognized fire as a natural process, and, increasingly, lightning ignitions were allowed to burn (Parsons 2000). But as these so-called natural fire policies moved from concept to practical application, questions and controversies arose (see Christensen 2005 for a review of these debates). Such policies suggested that fire be allowed to play a more natural role but failed to specify exactly what that might mean, leaving such decisions to managers at individual parks.

Sierra Nevada park scientists and managers, attempting to balance appropriate mixes of lightning and human ignitions and desirable fire frequencies, encountered several difficult questions (Graber 1983). One question revolved around whether a fire program needed to mimic aboriginal burning. In Sierran forests, for example, Kilgore and Taylor (1979) concluded that past fire frequencies were much greater than could be accounted for by lightning alone, suggesting a substantial role for Native American burning. This forced fire managers to ask whether it was sufficient to allow lightning ignitions or whether they needed to substitute for Native American ignitions as well.

Regarding desired outcomes of fire management policies, scientists and managers (Parsons et al. 1986: 21) proposed "that the principal aim of National Park Service resource management in natural areas is the unimpeded interaction of native ecosystem processes and structural elements." They asserted that this should be done by restoring fire as a natural process rather than using fire to restore particular ecosystem states or structures (Figure 5.1). Bonnicksen and Stone (1985) challenged this approach, arguing that naturalness cannot be restored by allowing fires to burn in forests with tree densities and other structural and compositional attributes that have been altered by years of fire suppression. According to these critics, active intervention is needed to restore ecosystem structure (thinning and perhaps even planting trees) before processes such as fire are restored. In larger parks, they noted, it may be possible to largely allow nature to take its course; however, in many parks ongoing active interventions, such as thinning forests and igniting prescribed burns, are likely to be necessary.

Vale (1987) points out that the argument over whether natural fire

FIGURE 5.1. Prescribed burning in the Giant Forest of Sequoia National Park. (Photo by Ted Young, National Park Service)

programs should focus on process or state and structure reflects differences in the degree of precision used in the definition of *natural* and the values attached to different ecosystem attributes. Indeed, the "process restorationists," with their less precise definition of *naturalness*, argue that it is inappropriate to define desired ecosystem states because too little is known about past conditions, there is too much emphasis on the single attribute of overstory structure, and it is too arbitrary to select a particular time in the past as a reference (Parsons et al. 1986). But others argue (perhaps for processes other than fire) that it is more difficult to develop process goals than structural goals, because processes are harder to measure and comprehend (Porter and Underwood 1999).

Lemons (1987) points out that although these questions are usually argued on scientific grounds, they are really questions of social values. With a policy based on the concept of naturalness, many of these questions cannot be answered by science alone. Are we concerned with process, structure, or both? What time period should be used as a basis for interventions? How much variability in conditions, due to chance and changing climate, should be allowed (Kilgore and Heinselman 1990)? Should decisions vary depending on whether the intervention is a one-time corrective measure or an ongoing management intrusion?

Despite efforts to restore fire to the large parks in the Sierra Nevada, fires still burn at less than 10 percent of historical frequencies (Caprio and Graber 2000). Often, "safer" surface fires are allowed to burn, whereas large, hot fires are suppressed, despite evidence that infrequent, large, intense fires were historically important to the structuring of ecosystems. So although improved scientific understanding of the ecological importance of fire and the dramatic fuel buildup across western forests spurred policies to allow lightning ignitions to burn in parks and wilderness, popular understanding of fire lagged behind science. Consequently, fire events such as the 1988 fires in Yellowstone were portrayed by the media and political leaders as ecological disasters and policy failures. According to Parsons (2000: 276), "Despite abundant evidence of the importance of fire as a natural process, and legislative and policy direction to preserve natural conditions (including the process of fire) in wilderness, fire suppression remains the dominant wilderness fire management strategy" because of administrative, political, and practical constraints, including concerns about human health and safety and damage to property and timber resources. Thus, like ungulate management, fire management has been influenced at least as much by social and political factors as by science.

The Problem of Broad and Ambiguous Policy

Debate over the management of ungulates and fire continues today, leaving these issues largely unresolved. Naturalness was once believed to provide important and much-needed guidance to managers regarding when and how to intervene to protect park and wilderness ecosystems. But, as demonstrated in the examples just described, policies directing managers to preserve or maintain natural conditions are vague and ambiguous and can be difficult to put into practice.

More recent agency policies, such as the National Park Service's Management Policies (2006), respond to both management challenges and recent ecological research but continue to leave some fundamental questions unanswered. The 2006 policies direct the Park Service to "maintain all the components and processes of naturally evolving park ecosystems, including the natural abundance, diversity, and genetic and ecological integrity of the plant and animal species native to those ecosystems" and recognize "natural change . . . as an integral part of the functioning of natural systems" (p. 36). Although these policies acknowledge ecosystem dynamism and the need to conserve biodiversity, they offer little specific guidance regarding where and when to intervene and how to define the desired outcomes. National

Park Service policies state that the goal of management is the preservation of all components and processes "in their natural condition," with natural condition defined as "the condition of resources that would occur in the absence of human dominance over the landscape" (p. 36). These policies also state that although intervention should be "kept to the minimum necessary," "biological or physical processes altered in the past by human activity may need to be actively managed to restore them to natural conditions or to maintain the closest approximation of the natural conditions when a truly natural system is no longer attainable" (p. 39).

Although the Wilderness Act is a more prescriptive statute than the National Park Service Organic Act, the Wilderness Act also fails to specify the meaning of naturalness. And like the National Park Service's policies, the Forest Service's Wilderness Management Policy (USDA Forest Service 2007) is broad and vague, directing managers to "maintain wilderness in such a manner that ecosystems are unaffected by human manipulation and influences so that plants and animals develop and respond to natural forces" (p. 7) and to "manage forest cover to retain the primeval character of the environment and to allow natural ecological processes to operate freely" (p. 38). In both cases, recent agency policies articulate broad, conceptual goals but, perhaps intentionally, provide little specific direction.

Lack of specificity is typical of natural resource and public land policy. Such policy usually provides a broad conceptual vision. The intention is that specific, actionable plans be developed by individual managers through local level planning. One consequence of this is that individual managers have much discretion to decide how to put the concept of naturalness into practice in a manner that takes into account local social and ecological conditions. The strength of this approach is that managers have the flexibility to tailor practices to fit the local context and respond to change. Potential weaknesses include inconsistency between adjacent units, shifting interpretations of policy caused by personnel changes, and lack of specificity in goals and targets. Cheever (1997: 638–639) suggests that "almost anything can be justified" because many of the statutes governing public lands are broad, offer little direction, and "speak of general values in mandatory terms." Nic (2004: 236) argues that this policy vacuum "has been filled with various agency interpretations and management philosophies."

The concept of naturalness is so broad and vague that a wide variety of policy interpretations and management actions can be pursued and justified. One Forest Service district ranger might decide that allowing a lightning fire to burn preserves a natural process, whereas an adjacent ranger might suppress such a fire to preserve a "primeval" old-growth forest. One park superintendent might decide to poison nonnative fish to restore historic aquatic communities, whereas another might decide that such an intervention is inappropriate. The resulting lack of coordination and consistency makes it difficult to achieve conservation goals at landscape and regional scales, the scales where many ecosystem processes operate (e.g., restoring fire or native predators is possible only across scales larger than individual parks or national forests). And as Lemons and Junker (1996: 394–395) point out, "ambiguity about the meaning of 'natural' has permitted wide oscillations in National Park Service policy regarding the acceptability of phenomena such as non-human caused fires, floods, and fluctuations in animal populations as well as of human intrusions in park ecosystems." If interpretations of policy fluctuate over time, long-term conservation objectives will be difficult to achieve. Conservation objectives will be better served by replacing the haphazard and random diversity in management practice with diversity that is planned and purposeful, a topic explored in more detail in Chapter 13.

The challenge of ambiguity is compounded by multiple mandates, both within and between protected area policies. Protected area policy expanded in the 1960s and 1970s to include a broader range of social values and political interests, but these new laws and regulations did little to clarify park purposes. Instead, new policies established multiple mandates that sent managers in ever-expanding and often divergent directions, with little guidance regarding how to make difficult decisions. Protected areas are now required to conserve imperiled species, protect air and water quality, provide recreation opportunities, and restore historic conditions. These potentially conflicting mandates are rarely prioritized at the agency, system, or unit level. An exception is the National Wildlife Refuge System Improvement Act (1997), which provides a hierarchy of goals and thus some direction regarding what managers should prioritize (Fischman 2002). Vague and conflicting policy can further politicize the decision-making process (Nie 2004) as agencies must constantly juggle political pressures, scientific uncertainty, and trade-offs between contradictory protected area values and goals. Thus, the administrative discretion that agencies have typically fought to preserve (Sellars 2000) can place them in a difficult position with the public. Managers must make highly political decisions with very little guidance regarding priorities.

Clear, unambiguous policy is critical to effective implementation of policy mandates (Nie 2004), and clear policy must start with well-articulated goals. Managers need to be able to translate policy goals into specific, attainable operational objectives. The need for more specific guidance increases with the complexity of conservation challenges. Responding effectively to climate change and other widespread stressors demands coordinated and thoughtful management strategies, implemented over large spatial scales.

This will not be possible until protected areas have addressed "the central question of what they will try to protect in an era of rapid climate change" (Lemieux et al. 2008: 12). Similarly, Kareiva et al. (2008) argue that the first step toward climate change adaptation is to clarify management goals and to determine whether goals should be adjusted.

Moving Toward More Specific Goals

Conservation of protected areas has evolved since the first national parks were established more than a hundred years ago. Native species preservation is now largely embraced, and sources of disturbance, such as fires and floods, are understood to play an integral ecological role. However, a variety of stressors threaten park and wilderness ecosystems. Such threats raise the specter of more management intervention, while policy guidance regarding management intervention continues to be vague and even contradictory. Lack of specificity and the competing meanings of naturalness make it difficult for managers to translate policy into practice. Therefore, we conclude that it is time for a transparent, public dialogue to carefully consider new guiding principles and management strategies, the role of intervention, and difficult trade-offs.

BOX 5.1. THE CHALLENGE OF PUTTING PARK AND WILDERNESS POLICY INTO PRACTICE

- Multiple and sometimes conflicting policy mandates now govern parks and wilderness. More recent statutory guidance includes the following:
    Wildness, expressed in the Wilderness Act as "untrammeled."
    Public involvement in agency decision making (National Environmental Policy Act).
    Endangered species conservation (Endangered Species Act).
- Increasingly, these park purposes are in conflict, making it necessary for managers to make decisions in the context of trade-offs as they attempt to decide whether, when, and how to intervene in ecosystems and ecological processes.
- Managers have struggled to put policies into practice because
    The naturalness concept is vague and ambiguous.
    Naturalness has multiple and conflicting meanings.
    Multiple mandates send managers in increasingly diverse directions with little sense of priorities.
- Management success could be improved by developing goals that are both more specific and diverse.

REFERENCES

Beyer, H. L., E. H. Merriell, N. Varley, and M. S. Boyce. 2007. Willow on Yellowstone's northern range: Evidence for a trophic cascade? *Ecological Applications* 17:1563–1571.

Bonnicksen, T. M., and E. C. Stone. 1985. Restoring naturalness to national parks. *Environmental Management* 9:479–486.

Caprio, A. C., and D. M. Graber. 2000. Returning fire to the mountains: Can we successfully restore the ecological role of pre-European fire regimes to the Sierra Nevada? Pp. 233–241 in D. N. Cole, S. F. McCool, W. T. Borrie, and J. O'Loughlin, comps. *Wilderness science in a time of change conference*. Vol. 5: *Wilderness ecosystems, threats, and management*. Proceedings RMRS-P-15-VOL-5. USDA Forest Service, Rocky Mountain Research Station, Ogden, UT.

Cheever, F. 1997. The United States Forest Service and National Park Service: Paradoxical mandates, powerful founders, and the rise and fall of agency discretion. *Denver University Law Review* 74:625–648.

Christensen, N. L. 1988. Succession and natural disturbance: Paradigms, problems, and preservation of natural ecosystems. Pp. 62–86 in J. K. Agee and D. R. Johnson, eds. *Ecosystem management for parks and wilderness*. University of Washington Press, Seattle.

Christensen, N. L. 2005. Fire in the parks: A case study for change management. *The George Wright Forum* 22(4):12–31.

Cole, D. N. 2000. Paradox of the primeval: Ecological restoration in wilderness. *Ecological Restoration* 18:77–86.

Fischman, R. L. 2002. The National Wildlife Refuge System and the hallmarks of modern organic legislation. *Ecological Law Quarterly* 29:457–622.

Graber, D. M. 1983. Rationalizing management of natural areas in national parks. *The George Wright Forum* 3:48–56.

Graber, D. M. 1995. Resolute biocentrism: The dilemma of wilderness in national parks. Pp. 123–135 in M. E. Soulé and G. Lease, eds. *Reinventing nature? Responses to postmodern deconstruction*. Island Press, Washington, DC.

Graber, D. M. 2003. Facing a new ecosystem paradigm for national parks. *Ecological Restoration* 21:264–268.

Huff, D. E., and J. D. Varley. 1999. Natural regulation in Yellowstone National Park's northern range. *Ecological Applications* 9:17–29.

Kareiva, P., C. Enquist, A. Johnson, S. H. Julius, J. Lawler, B. Petersen, L. Pitelka, R. Shaw, and J. M. West. 2008. Synthesis and conclusions. Pp. 9-1–9-66 in S. H. Julius and J. M. West, eds. *Preliminary review of adaptation options for climate-sensitive ecosystems and resources: A report by the U.S. Climate Change Science Program and the Subcommittee on Global Change Research*. U.S. Environmental Protection Agency, Washington, DC.

Kilgore, B. M., and M. L. Heinselman. 1990. Fire in wilderness ecosystems. Pp. 297–335 in J. C. Hendee, G. H. Stankey, and R. C. Lucas, eds. *Wilderness management*, 2nd ed. Fulcrum, Golden, CO.

Kilgore, B. M., and D. Taylor. 1979. Fire history of a sequoia–mixed conifer forest. *Ecology* 60:129–142.

Lemieux, C. J., D. J. Scott, R. G. Davis, and P. A. Gray. 2008. *Climate change, challenging choices: Ontario parks and climate change adaptation*. University of Waterloo, Department of Geography, Waterloo, ON.

Lemons, J. 1987. United States' national park management: Values, policy, and possible hints for others. *Environmental Conservation* 14:329–340.

Lemons, J., and K. Junker. 1996. The role of science and law in the protection of national park resources. Pp. 389–414 in R. G. Wright, ed. *National parks and protected areas: Their role in environmental protection*. Blackwell Science, Cambridge, MA.

Leopold, A. S., S. A. Cain, D. M. Cottam, I. N. Gabrielson, and T. L. Kimball. 1963. Wildlife management in the national parks. *American Forests* 69(4):32–35, 61–63.

Mann, C. C. 2005. *1491: New revelations of the Americas before Columbus*. Knopf, New York.

National Academy of Sciences. 2002. *Ecological dynamics on Yellowstone's northern range*. National Academy Press, Washington, DC.

National Park Service. 2006. *Management policies 2006*. Retrieved September 21, 2009 from www.nps.gov/policy/MP2006.pdf.

Nie, M. 2004. Statutory detail and administrative discretion in public lands governance: Arguments and alternatives. *Journal of Environmental Law & Litigation* 19:223–291.

Parsons, D. J. 2000. The challenge of restoring natural fire to wilderness. Pp. 276–282 in S. F. McCool, D. N. Cole, W. T. Borrie, and J. O'Loughlin, eds. *Wilderness science in a time of change conference*. Vol. 2: *Wilderness within the context of larger systems*. Proceedings RMRS-P-15-VOL-2. Rocky Mountain Research Station, Ogden, UT.

Parsons, D. J., D. M. Graber, J. K. Agee, and J. W. van Wagtendonk. 1986. Natural fire management in national parks. *Environmental Management* 10:21–24.

Porter, W. F., and H. B. Underwood. 1999. Of elephants and blind men: Deer management in the U.S. national parks. *Ecological Applications* 9:3–9.

Pritchard, J. A. 2002. The meaning of nature: Wilderness, wildlife, and ecological values in the national parks. *The George Wright Forum* 19(2):46–56.

Ripple, W. J., E. J. Larsen, R. A. Renkin, and D. W. Smith. 2001. Trophic cascades among wolves, elk and aspen on Yellowstone National Park's northern range. *Biological Conservation* 102:227–234.

Runte, A. 1979. *National parks: The American experience*. University of Nebraska Press, Lincoln.

Sellars, R. W. 1989. Science or scenery? A conflict of values in the national parks. *Wilderness* 52:29–38.

Sellars, R. W. 1997. *Preserving nature in national parks*. Yale University Press, New Haven, CT.

Sellars, R. W. 2000. The path not taken: National Park Service wilderness management. *The George Wright Forum* 17(4):4–8.

Singer, F. J., D. M. Swift, M. B. Coughenour, and J. D. Varley. 1998. Thunder on the Yellowstone revisited: An assessment of management of native ungulates by natural regulation, 1968–1993. *Wildlife Society Bulletin* 26:375–390.

Smith, D. W., R. O. Peterson, and D. B. Houston. 2003. Yellowstone after wolves. *BioScience* 53:330–340.

USDA Forest Service. 2007. Wilderness management. Pp. 1–55 in *Forest Service manual 2300: Recreation, wilderness, and related resource management*. USDA Forest Service, Washington, DC.

Vale, T. R. 1987. Vegetation change and park purposes in the high elevations of Yosemite National Park, California. *Annals of the Association of American Geographers* 77:1–18.

Wagner, F. H. 2006. *Yellowstone's destabilized ecosystem: Elk effects, science and policy conflicts*. Oxford University Press, Oxford.

Wagner, F. H., R. Foresta, R. B. Gill, D. R. McCullough, M. R. Pelton, W. F. Porter, and H. Salwasswer. 1995. *Wildlife policies in the U.S. national parks*. Island Press, Washington, DC.

Wright, R. G. 1999. Wildlife management in the national parks: Questions in search of answers. *Ecological Applications* 9:30–36.

PART II

Approaches to Guide Protected Area Conservation

Earlier chapters emphasized the need to move beyond the concept of naturalness to provide guidance to park and wilderness managers faced with deciding where, when, and how to intervene in ecosystems. Although naturalness provides a foundation on which to build stewardship goals and objectives, more specific and diverse goals are needed.

In the second part of this book, chapter authors explore four goals or management emphases, each founded in but going beyond traditional notions of naturalness. The four goals described here do not capture all potential management emphases, but they provide a range of approaches and include both traditional emphases, such as hands-off management and historical fidelity, and more recent emphases, such as ecological integrity and resilience. Each of these emphases, like naturalness, can mean different things to different people. So definitions are important. The interpretations presented by chapter authors may not capture all possible ways these emphases might be construed.

There is substantial overlap between these emphases; they are not mutually exclusive. In each chapter, authors articulate the more unique aspects of each emphasis. But in many cases, the appropriate management approach might be to seek a balance between several emphases, working in the overlap between them. This brings us to one of the most important theses of this book: Pursuing different goals and objectives in different places should optimize the sustainability of all park and wilderness values into the future. Therefore, although each emphasis is presented as a separate chapter, they should be thought of as working together—as an array of approaches to be thoughtfully and purposefully applied in a diverse way across large landscapes and systems of protected areas—a topic returned to in the third part of this book.

The desired outcome of the approach described in Chapter 6 is autonomous nature: ecosystems that are wild, self-willed, and untrammeled. This is the emphasis most consistent with the meaning of *naturalness* that focuses on freedom from intentional human control. The means of achieving this outcome are clear: Park and wilderness managers should not manipulate or intervene in ecosystem processes for any purpose, even to restore lands degraded by human activity. They should simply let them be.

In contrast, pursuing the emphasis presented in Chapter 7, ecological integrity, often entails substantial intervention. Active management is needed in many places to maintain the integrity of park and wilderness ecosystems assaulted by changes that threaten biodiversity. Like *naturalness*, *ecological integrity* is a vague and ambiguous term until it is defined. In contrast to *naturalness*, *ecological integrity* shifts the focus of attention from the past to the future and from human cause to ecological effect. Once defined, *ecological integrity* allows the specification of explicit objectives and measurable endpoints. Chapter 7 explores Parks Canada's abandonment of naturalness and adoption of ecological integrity as their framework for ecosystem stewardship.

At one time it was widely believed that the remaining two emphases, historical fidelity and resilience, went hand in hand. Because historical ecosystems had evolved in place over long time periods, in conjunction with historical disturbance regimes, the recipe for resilience was to keep systems within the bounds of their historical range of variability. But with climate change, that premise is no longer tenable. A gap has developed between an emphasis on historical fidelity and an emphasis on resilience, and tension is building.

Historical fidelity, the subject of Chapter 8, is the emphasis most concerned with maintaining compositional and structural legacies. Enhancing the resilience of park and wilderness ecosystems, as explored in Chapter 9, involves encouraging adaptability, embracing some change in order to avoid catastrophic change. The emphasis is more on function than on composition and structure. When emphasizing resilience, managers must define which ecosystem elements they want to be resilient and what they want them to be resilient to. As with all these approaches, choices must be made—choices partially informed by science but ultimately driven by human values.

Chapter 6

Let It Be: A Hands-Off Approach to Preserving Wildness in Protected Areas

PETER LANDRES

We should have the wisdom to know when to leave a place alone.
—*Sir Peter Scott*

In an era of rapid global climate change and other pervasive anthropogenic ecological insults, many scientists and managers have few qualms about taking action to mitigate the effects of these insults, including in areas that are protected by law as wilderness, wildlife refuges, or national parks. For example, habitat is manipulated to sustain populations of selected threatened and endangered species, nonindigenous invasive species are removed, and extirpated species are reintroduced. Even with important technical advances in the ecological sciences over the last several decades, ecologists still question the feasibility of managing biodiversity in the face of continued environmental change. For example, Western (2004: 496) wryly notes in an essay on the paradox of managing wildlands that "like the Red Queen running in place, we are destined to manage ever harder to save any semblance of the natural until . . . the unmanaged will be more managed than the managed to preserve the illusion of the natural."

Is there an alternative to becoming locked into managing what is not fully understood in an environmental context that is rapidly changing? The purpose of this chapter is to explore the reasoning behind and steps toward implementing the hands-off alternative. This alternative focuses on one of the three meanings of *naturalness* described in Chapter 2: freedom from intentional human control, intervention, and manipulation. This chapter first reviews current understandings about wildness and the autonomy of nature as the context for a hands-off approach. Next, the reason for leaving some areas alone, expressly because of global climate change and other novel ecological changes that are occurring, and the substantial and unique benefits to the land and people from this approach are discussed. Last, limitations and the specific conditions under which such an approach might be most feasible are explored. I conclude that although the hands-off alternative is not feasible in all areas, in some areas it is, and that more fully understanding what is gained and what is lost by such an approach fosters the opportunity to realize all the meanings and benefits of naturalness in the twenty-first century.

Defining Wildness

In his "Walking" essay, Thoreau (1862: 664) wrote, "In wildness is the preservation of the world." Since then, many terms and phrases have been used to describe "wild" as self-willed, autonomous, unmanipulated, unrestrained, uncontrolled, unbounded, unimpeded, and free. Turner (1996: 112) writes that "a place is wild when its order is created according to its own principles of organization—when it is self-willed land." Schroeder (1994: 64) suggests that "our responsibility . . . is to respect the autonomy of nature—to care about nonhuman nature for its own sake and grant it at least a measure of freedom to follow its own path." In *Recognizing the Autonomy of Nature*, Heyd (2005) points out that *autonomy* is based on the root words *autos* and *nomos*, literally meaning "self-rule." In the context of environmental conservation, the autonomy of nature is generally considered to mean that an area, ecosystem, species, or ecological process follows its own internal drives (see Katz 1997). Hettinger (2005: 90) states that "nature carries on independently of human control or domination" and that nature's autonomy is respected by "avoiding exerting" human influence over it.

Examining how this concept of autonomy applies to ecological restoration in wilderness, Woods (2005) argues that there are really two distinct embedded concepts: wildness and freedom. Wildness is the lack of intentional manipulation that allows self-expression, whereas freedom is the lack of external constraints that limit the capacity for this self-expression. In other words, an area is wild when it is not intentionally manipulated, even though its freedom may still be reduced because global climate change is altering temperature and precipitation regimes (the external context).

Woods (2005) further elaborates on the distinction between wildness and naturalness, echoed in the discussion in Chapter 2. This conceptualization of wildness and how it is related to but distinct from freedom and naturalness helps clarify what have been confusing or unfounded management implications (Ridder 2007), such as that " 'hands-on' management is needed to restore 'hands-off' wilderness character" (Noss 1985: 19). For example, spraying herbicides compromises wildness but was considered necessary to protect the indigenous plants in the Frank Church River of No Return Wilderness (Anderson and Wotring 2001). This recent discussion of wildness can also "be seen as a sign of a growing sensitivity towards the meaning of nature, an emerging new 'wildness ethic' " (Drenthen 2007: 394). Similarly, Willers (1999: 3) concludes, "Reawakening to wildness will carry with it deference to the dense meshwork of process that has been the source of evolution since life on Earth began, and an acknowledgment of the inherent rightness of allowing some vast landscapes to function, with all parts intact, according to their own internal dictates." As a scientific and moral concept, wildness may therefore help society more fully understand, respect, and value nature's autonomy.

The Role of Designated Wilderness

As a legal mandate and policy goal, no area is explicitly protected for wildness, although in the United States wilderness designated under the 1964 Wilderness Act (Public Law 88-577) comes closest. In defining *wilderness*, this law states, "A wilderness, in contrast with those areas where man and his own works dominate the landscape, is hereby recognized as an area where the earth and its community of life are untrammeled by man." The word *untrammeled* is rarely used, but Howard Zahniser, the primary author of the Wilderness Act, purposefully used this word as the key element in the definition of *wilderness* (Scott 2002). The intent behind using the word *untrammeled* was to establish a relationship between people and the land that was based on restraint, humility, and respect—a relationship that would foster a sense of interdependence and interconnectedness that people felt with the land (Zahniser 1956). Olson (1976) and many others argue that the loss of this relationship between people and the land is a root cause for many of the conservation problems seen today.

As Friskics (2008) points out, recent criticism of wilderness misconstrues *untrammeled* to mean pristine, unaffected by indigenous people, or not influenced by exogenous environmental threats. Instead, in the management context, *untrammeled* means that "wilderness is essentially unhindered and free from modern human control or manipulation" (Landres et al. 2008: 7). This idea builds on the simple description by Zahniser (1963: 2) that wilderness managers are "guardians not gardeners." This implies that to maximize the untrammeled quality in wilderness, managers should refrain from taking actions that manipulate, control, or intervene with the ecological system.

A Hands-Off Approach

Although humility and restraint should be hallmarks for the management of all protected areas, as well as the other approaches that are described in this section, the hands-off approach takes restraint to an extreme. The hands-off approach, in its simplest form, is not taking action that manipulates, controls, or hinders the conditions (e.g., habitat), components (e.g., species), or processes (e.g., fire) of an ecological system. This restraint should be based not on naiveté or wishful thinking that there is not an ecological problem, or on a lack of understanding about the consequences and trade-offs of restraint, but rather on a purposeful and willful holding back to respect nature's autonomy and to observe and learn from what happens. In contrast to the other chapters in this section that advocate management intervention to sustain elements of historic fidelity, resilience, or ecological integrity, the goal of the hands-off approach is to not intervene even if some of these elements are lost. As Lucas (1973: 151) suggests, in wilderness "the object is not to stop change, nor to recreate conditions as of some arbitrary historical date, nor to strive for favorable change in big game populations or in scenic vistas. The object is to let nature 'roll the dice' and accept the results with interest and scientific curiosity."

The hands-off approach preserves wildness by restraining direct interventions generally recognized as having adverse effects on ecosystems, such as suppressing lightning-ignited fires, introducing nonindigenous fish and ungulates for sport (or any other reason), killing or removing native predators, or damming and diverting water bodies and flows. In addition, and much harder for ecologists and conservationists to accept, the hands-off approach restrains interventions taken to provide specific eco-logical benefits, such as using management-ignited prescribed fire to reduce accumulated forest fuels and mimic a natural fire regime, spraying herbi-cides or introducing biological control agents to eradicate nonindigenous invasive plants, or creating artificial water or food sources to replace those that are no longer accessible because of habitat fragmentation.

Factors Affecting Whether a Hands-Off Approach Is Appropriate

Three intertwined factors affect whether a hands-off approach is appropri-ate and will be used: legislation, ethics, and knowledge. First, the legislative mandate for an area fundamentally determines whether management inter-vention is warranted. For example, in areas that are designated for the pro-tection of threatened or endangered species, such as some national wildlife refuges, active management interventions are usually mandatory to support the persistence of such species. In contrast, in designated wilderness and potentially in other protected areas that have similar legal direction, such as Adirondack Park in New York State, the hands-off approach would be con-sidered appropriate. However, some areas are managed under overlapping legislative mandates, such as Cabeza Prieta Wilderness in Arizona, which is managed as wilderness and for the endangered Sonoran pronghorn ante-lope (*Antilocapra americana sonoriensis*). In such cases the appropriateness of a hands-off approach must be determined by the specifics of the case and area.

Ethics, based on the values of individual managers, scientists, and pub-lic stakeholders, may influence the interpretation of uncertain or ambigu-ous legislation (where not already determined by judicial opinions) and typically has a strong impact on whether the hands-off approach is con-sidered appropriate. Katz (1992) argues that managing "natural entities" and attempting the "technological fix of nature" are basically moral and value-laden decisions. For example, the decision about whether the hands-off approach is ethically appropriate may rest on values-based questions such as the following:

- Should management interventions be used to sustain specific ele-ments of biodiversity or ecological processes in wilderness or simi-larly protected areas?
- Should precaution dominate and no action be taken unless no harm can be demonstrated from action-based alternatives?

- How does a manager decide whether preserving wildness (hands off) or biodiversity (management intervention) has priority?
- Under what circumstances is it appropriate to set a precedent for interventions that compromise wildness?

The last major factor affecting the use of a hands-off approach is whether sufficient knowledge exists to manage an ecosystem or specific elements within an ecosystem. One reason for restraint in general and the hands-off approach in particular is that in many situations there is insufficient knowledge to manage ecosystems. Any environmental science textbook is replete with examples of attempts to "fix" an ecological problem that went awry because of unforeseen consequences. Global climate change exacerbates this lack of knowledge. In discussing whether there is sufficient knowledge to manage ecosystems, Turner (1996: 124) concludes, "We are not that wise, nor can we be. The issue is not the legitimacy of science in general, nor the legitimacy of a particular scientific discipline, but the appropriate limits to be placed on any scientific discipline in light of limited knowledge. To ignore these limits is to refuse humility." The increasing rarity of wildness in our increasingly manipulated world argues for greater humility and restraint, for watching change occur—even if this change is not in accord with what managers and scientists think should be happening.

The Benefits of a Hands-Off Approach

Wilderness, national parks, and other areas that are designated by law or administrative policy are generally protected for their social and ecological values. In this section I describe some of the social and ecological benefits of a hands-off approach.

DEEPENING RESPECT FOR NATURE'S AUTONOMY

An important benefit of the hands-off approach is a deepening awareness of and respect for nature's autonomy. This is a long-term societal value and benefit that is at the heart of the direct interplay between people and their environment (Plumwood 2005). The hands-off approach is a conscious choice to put restraint first, to ensure that people are not in charge, in control, or dominating, to foster awareness of and appreciation for our interconnectedness with what is typically called nature. Rolston (1999), Schroeder (2007), and Keeling (2007) respectively describe the significant spiritual, psychological, and philosophical benefits to individuals and to the larger society from having areas where people are purposefully not in control. In other words, not intentionally manipulating or controlling offers an antidote to some of the deeper spiritual, psychological, and philosophical problems caused by our increasingly developed and manipulated world. To the extent that the value of this approach is recognized and articulated, our society has the opportunity to move toward a deeper and more enduring relationship with all three meanings of *naturalness* described in Chapter 2.

FOSTERING SCIENTIFIC HUMILITY

A derivative benefit of such restraint is the opportunity for scientists to be placed in a position where they explicitly acknowledge the limitations of their understanding about ecological systems; that is, where they may gain humility. In a recent meta-analysis of ecological surprises, Doak et al. (2008) contend that "the extent and frequency of major 'surprises' in ecological systems argue for substantial humility about our predictive abilities" (p. 953). Humility is needed, they say, because scientists are "sometimes surprised because of ignorance, sometimes because of a failure to pay careful attention, and sometimes because [they] have to prioritize which aspects of ecology to include and which to ignore in order to make predictions" (p. 957). They also assert that "most management strategies, sooner or later, will not work as planned . . . sometimes not just less than perfect in achieving some desired outcome, but totally wrong" and that "frequent ecological surprises reinforce the need for management plans that are highly precautionary" (p. 958). As a specific example, in an experimental study to elucidate patterns of plant zonation in Chilean salt marshes, Farina et al. (2009) found that the long-established mechanistic understandings of pattern generation in salt marshes derived from California and New England could not be exclusively used to manage and restore the novel Chilean salt marsh systems.

ACCEPTING EVOLUTIONARY CHANGE

There are likely to be ecological benefits from a hands-off approach, although there are few direct experimental studies to support the ideas explored here. First, this approach might increase the likelihood of evolutionary processes being largely unfettered by modern people. Ashley et al. (2003), Bøhn and Amundsen (2004), and Klein et al. (2009) discuss the importance of explicitly considering evolutionary outcomes in conservation strategies. Genetic frequencies will always change in response to whatever selection pressures are extant in the area, so evolution per se will continue unabated no matter how much modern people intervene. Site-specific and pervasive ecological changes, from global climate change and airborne pollutants to increasing human presence and development, affect selection pressures everywhere and consequently evolutionary outcomes. Western (2004: 496) comments that "the more we change the world, the more we govern evolution."

At stake are selection pressures and evolutionary outcomes that are intentionally not dominated or controlled by the whims, desires, designs, and even good intentions of modern people. For example, in a study of selection pressures among native and invasive perennial and annual grasses, Leger (2008: 1226) concludes that "while it is tempting to restore degraded areas to higher densities of natives . . . such actions may impede long-term adaptation to new conditions by arresting or reversing the direction of ongoing natural selection in the resident population." At its core, the hands-off approach strives to protect and sustain the types and intensities of selection pressures that gave rise to the diversity and complexity of life seen today and to allow these forces to operate in ways that people do not intentionally control and could not fathom in the future. Respecting nature's autonomy is allowing evolutionary change and adaptation to occur, even in response to what are typically considered environmental insults

SUSTAINING NONFOCAL SPECIES

Another ecological benefit of the hands-off approach is that it may increase the likelihood of protecting a broad range of species that might otherwise be lost when management focuses on select species. For example, Ozaki et al. (2006) found that management plans for the northern goshawk (*Accipiter gentilis*), often cited as an umbrella for other species, failed to protect the diversity of birds, butterflies, carabid beetles, and forest floor plants in Japan that use the same habitat as the goshawk. More generally, in a recent review of ecological restoration goals, Choi et al. (2008: 60) conclude that "restoration goals are determined by us, not by nature. . . . For this reason, the goals tend to be determined by preconceptions or misconceptions that often place more value on certain target species or ecosystems." By definition, the hands-off approach has no a priori ecological target, focal species, or ecological process. Instead, the intent is to let all species and ecological processes in the area be or change, free from intentional manipulation, and not focus management on any particular component or process.

REDUCING UNINTENDED ADVERSE CONSEQUENCES

Another potential ecological benefit of a hands-off approach is that scientists and managers would not inadvertently cause adverse impacts to ecosystems when trying to help. For example, in a review of the effects of management interventions to improve the resistance and resilience of U.S. northeastern forests in advance of insect infestation and disease, Foster and Orwig (2006: 968) conclude, "Current management regimes aiming to increase long-term forest health and water quality are ongoing 'experiments' lacking controls. In many situations good evidence from true experiments and 'natural experiments' suggests that the best management approach is to do nothing." Similarly, in experiments testing the use of herbicides to reduce the abundance of invasive plants and their impact on native species, Rinella et al. (2009: 155) found that herbicide use made two native plant species "exceedingly rare" over the 16-year study and that the "dominant invader became more abundant in response to the decreases in native-forb abundance."

PROVIDING UNMANIPULATED BENCHMARKS

A hands-off approach increases the likelihood of an area serving as an unmanipulated ecological benchmark, or at least as unmanipulated as may be possible in the context of an increasingly humanized landscape and planet. For example, Van Mantgem et al. (2009) located sixty-six undisturbed old-growth forest stands to examine the causes of tree mortality across the western United States. This may be closest to what Leopold (1941: 3) described as "the base datum of normality, a picture of how healthy land maintains itself as an organism." Such a benchmark does not need to be based on an arbitrary timeframe (such as pre-European settlement) or on whether indigenous people influenced the area. Instead, the area is simply not intentionally manipulated or controlled from the time and state of designation onward.

PRESERVING OPTIONS AND HEDGING RISK

Finally, a hands-off approach preserves options for the future. Given huge uncertainty about the type and intensity of both current and future stressors that will adversely affect ecological systems, especially global climate change, and huge uncertainty about the response of ecological systems to these stressors (Hobbs et al. 2006), the more options for future responses and management the better (Seastedt et al. 2008). The hands-off approach contributes a unique perspective to the suite of management options and possibilities, thereby keeping as many options open as possible in the coming time of rapid and novel ecological change.

Limitations and Barriers to a Hands-Off Approach

A hands-off approach is still a form of management, and like all management it entails limitations and trade-offs. The most important limitation posed by the hands-off approach is the increased risk to specific elements of biodiversity (Figure 6.1). For example, if decades of fire exclusion have

FIGURE 6.1. Whitebark pine (*Pinus albicaulis*) illustrates the dilemma of deciding whether to use a hands-off approach. It occurs throughout the Pacific Northwest and northern Rocky Mountains, and in certain seasons its seeds form a significant part of the diet of endangered grizzly bears. The pine is severely declining in many parts of its range because of fire exclusion, attack by the nonindigenous whitepine blister rust (*Cronartium ribicola*), and increasing infestations of mountain pine beetle (*Dendroctonous ponderosae*), probably allowed by climate change. Several proactive treatments being proposed to restore pine populations include replanting forests with seed gathered from trees that are naturally resistant to the blister rust. The main thesis of this chapter is that the goal of preserving wildness via a hands-off approach should be considered for some areas, even though it may allow the decline of whitebark pine in these areas. (Photo by John Schwandt, Forest Service)

allowed current fuel loads to build, a naturally ignited fire that is allowed to burn may cause the death of old-growth ponderosa pines (*Pinus ponderosa*), which are of high social and ecological value (Allen et al. 2002). Conversely, in the absence of active efforts to restore the fire regime with the use of management-prescribed fires, some plant and animal species might decline or become locally extinct (Christensen 1988; Agee 2002). Likewise, nonindigenous invasive plants may cause the loss or reduced distribution of indigenous species (Orrock et al. 2008), and aggressive eradication may be the only way to sustain native species (see Anderson and Wotring 2001 for an example in wilderness). Comparing the effects of active management and no management of forests to sustain red-cockaded woodpeckers (*Picoides borealis*), Saenz et al. (2001) found that using prescribed fire and creating artificial nest cavities was necessary for woodpecker persistence and that eight of nine woodpecker groups were lost in hands-off areas over a period of nearly 20 years. Graber (2003: 38) has consistently argued that some wilderness areas "require urgent intervention and long-term maintenance simply to preserve what remains—and often what remains is quite irreplaceable. To put it another way, their value as managed reserves of biodiversity exceeds their value as 'wilderness.' "

Cultural resources may also be put at greater risk with a hands-off approach. Sydoriak et al. (2001) describe how not taking intensive restoration actions to cut encroaching trees and replant grasses and other ground vegetation inside the Bandelier Wilderness will cause the loss of archaeological remains. Where specific resources, such as threatened or endangered species or cultural and heritage sites, are at risk and protected by legislation, the hands-off approach may not be appropriate.

A potential barrier to implementing the hands-off approach is that it may allow certain species or disturbance processes that pose a risk to life or property to spread outside the protected area. For example, because the boundaries of protected areas are porous, naturally ignited fire that is not suppressed, nonindigenous invasive plants that are not eradicated, or certain species such as bison or wolves that are allowed to thrive could spread outside the area managed under a hands-off approach. In such cases, public or management pressures may be sufficient to preclude implementing a hands-off approach.

Another significant barrier to implementing the hands-off approach is that it goes against the dominant paradigm in most management agencies that doing something is better than nothing and that managing by command and control is appropriate (Holling and Meffe 1996). Scientists may exhibit a similar attitude based on faith in their technical knowledge.

For example, Janzen (1998: 1312) contends that humanity must accept its responsibility for "recognizing and relabeling wildland nature as a garden per se, having nearly all the traits that we have long bestowed on a garden—care, planning, investment, zoning, insurance, fine-tuning, research, and premeditated harvest." Both managers and scientists may strongly desire to do good and think that today's rapid anthropogenic change brings a responsibility to stave off the loss of biodiversity. Adding to this belief, both groups may be unwilling to take the risk and liability of not doing something even if the technical knowledge is incomplete or uncertain. Although they are understandable, implementing the hands-off approach will require great conviction to overcome these attitudes.

Implementing a Hands-Off Approach

As part of a diverse suite of management approaches, the hands-off approach would be feasible only in certain areas. Discussing feasibility of the hands-off approach helps develop what Willers (1992: 605) calls "a science of letting things be" and is critical because if this approach remains only a naive ideal, then the societal and ecological benefits of wildness will never be realized. This discussion focuses on the intentional implementation of a hands-off approach, not a de facto hands-off approach that occurs in many remote areas, whether protected or not, because of a lack of staff and funding resources.

A hands-off approach would be least appropriate in areas where active management is needed for the protection and maintenance of species or communities, especially those that are listed as threatened, endangered, or sensitive. In addition, where the practices and traditions of indigenous people are an integral part of the ecological system, the hands-off approach probably would not be appropriate.

The hands-off approach is more feasible in areas that have legal and administrative policy direction that supports the goal of wildness. As already discussed, *nature's autonomy* and *wildness* are not explicitly used in law or policy for any U.S. agency or protected area. However, all designated wilderness broadly fits the goal of wildness because of the emphasis on untrammeled nature. Furthermore, the mandate of untrammeled nature requires only that the area be free from intentional manipulation, not that the area be free from human influence.

The hands-off approach is also more feasible in an area that is large and isolated. A large area provides the variety of terrain, ecological processes, disturbances, and resources sufficient to allow persistence of species' metapopulations and to fulfill the dispersal needs of a species (Noon and Dale 2002). Isolation provides buffering from threats that move short or moderate distances, such as many nonindigenous plant and animal species, and from the myriad other effects of fragmentation. Isolation also separates the area from the socioeconomic interests of people because these interests have an effect far beyond the area of personal or commercial property. For example, a naturally ignited fire that occurs inside a protected area may be suppressed because of the concern that it will spread outside and harm people or their property. Likewise, a naturally ignited fire outside a protected area may be suppressed for the same reasons, even though under more natural circumstances this fire might burn into the protected area.

Because isolation from roads or other developments is almost impossible to find in the United States (Watts et al. 2007), relative isolation will have to suffice. Such relative isolation could be achieved by using areas that are embedded in the center or core of a moderate to large area and ensuring that the management goals in the outlying areas are compatible with and provide a buffer to the core area (Landres et al. 1998). However, isolation will not protect any area from regional threats such as air pollution or from the effects of global climate change.

Finally, the hands-off approach would be most feasible as part of an integrated, diverse system of protected areas, each with specific goals that in combination provide sufficient protection and the benefits (both ecological and societal) that come from such broad-scale protection (Lambeck and Hobbs 2002). For example, a national park that is 90 percent designated wilderness could have a core area (depending on its spatial configuration) allocated to wildness that is managed with a hands-off approach. The wilderness outside this core would be managed more intensively to eradicate or control specific threats, such as nonindigenous invasive plants. And the area outside the wilderness but still within the park would be managed even more intensively to achieve specific management goals and help buffer the inner areas. This model of complementary management goals could also be applied across a landscape. For example, an endangered and endemic species that needs periodic habitat manipulation, such as low-intensity fire, could be protected in one area while another part of this landscape could be devoted to preserving wildness using a hands-off approach. In twenty-two 1-hectare experimental sites, Franc and Gotmark (2008) found that the number of beetle species increased with a combination of active and hands-off management: Red-listed beetle species declined in active management areas, whereas hands-off areas had more open patches and more dead wood that favored other species. Cole (2001) suggests that a combination of active management and hands-off approaches may be inevitable in wilderness.

Shall We Accept the Wild?

By willfully not manipulating or intervening in ecological systems, the hands-off approach is a way to foster greater respect and humility toward the autonomy of nature, to "let being be" (Abbey 1984: 43). Global climate change highlights how little managers and scientists understand about ecological systems, and respecting nature's autonomy and using a hands-off approach is even more important in such a novel world to hedge risk and not cause inadvertent problems. This approach will be difficult to implement given the impulses, the desires, and the burden of responsibility managers and scientists may strongly feel to help protect parks and wilderness in the twenty-first century. To truly protect these areas in an uncertain future, managers and scientists need to face what Turner (1996: 125) describes as a fundamentally moral choice: "Shall we remake nature according to biological theory? Shall we accept the wild?"

BOX 6.1. MANAGING FOR AUTONOMOUS NATURE

- Wildness is the lack of intentional management interventions that allow self-expression of the ecological system in an area.
- Designated wilderness, with its emphasis on untrammeled nature, comes closest to providing legal protection for wildness.
- The hands-off approach takes restraint to an extreme by not taking actions that manipulate, control, or hinder the ecological system in an area.
- The benefits of a hands-off approach may include deepening respect for nature's autonomy, fostering scientific humility, accepting evolutionary change, sustaining nonfocal species, reducing unintended adverse consequences, providing unmanipulated benchmarks, preserving options, and hedging risk.
- The hands-off approach poses increased risk to specific elements of biodiversity or other resources of high social value that need management interventions to sustain them.
- The hands-off approach would be most feasible in large, isolated areas that are part of a landscape-scale, integrated suite of conservation strategies.

REFERENCES

Abbey, E. 1984. *Beyond the wall: Essays from the outside*. Henry Holt, New York.

Agee, J. K. 2002. The fallacy of passive management. *Conservation Biology in Practice* 3:18–25.

Allen, C. D., M. Savage, D. A. Falk, K. F. Suckling, T. W. Swetnam, T. Schulke, P. B. Stacey, P. Morgan, M. Hoffman, and J. T. Klingel. 2002. Ecological restoration of southwestern ponderosa pine ecosystems: A broad perspective. *Ecological Applications* 12:1418–1433.

Anderson, B., and K. Wotring. 2001. Invasive plant management along wild rivers: Are we stewards, guardians, or gardeners? *International Journal of Wilderness* 7(1):25–29.

Ashley, M. V., M. F. Willson, O. R. W. Pergams, D. J. O'Dowd, S. M. Gende, and J. S. Brown. 2003. Evolutionarily enlightened management. *Biological Conservation* 111:115–123.

Bøhn, T., and P. A. Amundsen. 2004. Ecological interactions and evolution: Forgotten parts of biodiversity? *BioScience* 54:804–805.

Choi, Y. D., V. M. Temperton, E. B. Allen, A. P. Grootjans, M. Halassy, R. J. Hobbs, M. A. Naeth, and K. Torok. 2008. Ecological restoration for future sustainability in a changing environment. *Ecoscience* 15:53–64.

Christensen, N. L. 1988. Succession and natural disturbance: Paradigms, problems, and preservation of natural ecosystems. Pp. 62–86 in J. K. Agee and D. R. Johnson, eds. *Ecosystem management for parks and wilderness*. University of Washington Press, Seattle.

Cole, D. N. 2001. Management dilemmas that will shape wilderness in the 21st century. *Journal of Forestry* 99:4–8.

Doak, D. F., J. A. Estes, B. S. Halpern, U. Jacob, D. R. Lindberg, J. Lovvorn, D. H. Monson, et al. 2008. Understanding and predicting ecological dynamics: Are more surprises inevitable? *Ecology* 89:952–961.

Drenthen, M. 2007. New wilderness landscapes as moral criticism: A Nietzschean perspective on our contemporary fascination with wildness. *Ethical Perspectives* 14:371–403.

Farina, J. M., B. R. Silliman, and M. D. Bertness. 2009. Can conservation biologists rely on established community structure rules to manage novel systems? . . . Not in salt marshes. *Ecological Applications* 19:413–422.

Foster, D. R., and D. A. Orwig. 2006. Preemptive and salvage harvesting of New England forests: When doing nothing is a viable alternative. *Conservation Biology* 20:959–970.

Franc, N., and F. Gotmark. 2008. Openness in management: Hands-off vs. partial cutting in conservation forests, and the response of beetles. *Biological Conservation* 141:2310–2321.

Friskics, S. 2008. The twofold myth of pristine wilderness: Misreading the Wilderness Act in terms of purity. *Environmental Ethics* 30:381–399.

Graber, D. M. 2003. Ecological restoration in wilderness: Natural versus wild in National Park Service wilderness. *The George Wright Forum* 20(3):34–41.

Hettinger, N. 2005. Respecting nature's autonomy in relationship with humanity. Pp. 86–98 in T. Heyd, ed. *Recognizing the autonomy of nature: Theory and practice*. Columbia University Press, New York.

Heyd, T., ed. 2005. *Recognizing the autonomy of nature: Theory and practice*. Columbia University Press, New York.

Hobbs, R. J., S. Arico, J. Aronson, J. S. Baron, P. Bridgewater, V. A. Cramer, P. R. Epstein, et al. 2006. Novel ecosystems: Theoretical and management aspects of the new ecological world order. *Global Ecology and Biogeography* 15:1–7.

Holling, C. S., and G. K. Meffe. 1996. Command and control and the pathology of natural resource management. *Conservation Biology* 10:328–337.

Janzen, D. 1998. Gardenification of wildland nature and the human footprint. *Science* 279:1312–1313.

Katz, E. 1992. The call of the wild: The struggle against human domination and the technological fix of nature. *Environmental Ethics* 14:265–273.

Katz, E. 1997. *Nature as subject: Human obligation and natural community*. Rowman and Littlefield, Lanham, MD.

Keeling, P. K. 2007. Beyond the symbolic value of wildness. *International Journal of Wilderness* 13(1):19–23.

Klein, C., K. Wilson, M. Watts, J. Stein, S. Berry, J. Carwardine, M. S. Smith, B. Mackey, and H. Possingham. 2009. Incorporating ecological and evolutionary process into continental-scale conservation planning. *Ecological Applications* 19:206–217.

Lambeck, R. J., and R. J. Hobbs. 2002. Landscape and regional planning for conservation: Issues and practicalities. Pp. 360–380 in K. J. Gutzwiller, ed. *Applying landscape ecology in biological conservation*. Springer-Verlag, New York.

Landres, P., C. Barns, J. G. Dennis, T. Devine, P. Geissler, C. S. McCasland, L. Merigliano, J. Seastrand, and R. Swain. 2008. *Keeping it wild: An interagency strategy to monitor trends in wilderness character across the National Wilderness Preservation System*. General technical report RMRS-GTR-212. USDA Forest Service Rocky Mountain Research Station, Fort Collins, CO.

Landres, P., S. Marsh, L. Merigliano, D. Ritter, and A. Norman. 1998. Boundary effects on wilderness and other natural areas. Pp. 117–139 in R. L. Knight and P. B. Landres, eds. *Stewardship across boundaries*. Island Press, Washington, DC.

Leger, E. A. 2008. The adaptive value of remnant native plants in invaded communities: An example from the Great Basin. *Ecological Applications* 18:1226–1235.

Leopold, A. 1941. Wilderness as a land laboratory. *The Living Wilderness* 6(July):3.

Lucas, R. C. 1973. Wilderness: A management framework. *Journal of Soil and Water Conservation* 28:150–154.

Noon, B. R., and V. H. Dale. 2002. Broad-scale ecological science and its application. Pp. 34–52 in K. J. Gutzwiller, ed. *Applying landscape ecology in biological conservation*. Springer-Verlag, New York.

Noss, R. F. 1985. Wilderness recovery and ecological restoration: An example for Florida. *Earth First!* 5(8):18–19.

Olson, S. F. 1976. *Reflections from the north country*. Knopf, New York.

Orrock, J. L., M. S. Witter, and O. J. Reichman. 2008. Apparent competition with an exotic plant reduces native plant establishment. *Ecology* 89:1168–1174.

Ozaki, K., M. Isono, T. Kawahara, S. Iida, T. Kudo, and K. Fukuyama. 2006. A mechanistic approach to evaluation of umbrella species as conservation surrogates. *Conservation Biology* 20:1507–1515.

Plumwood, V. 2005. Toward a progressive naturalism. Pp. 25–53 in T. Heyd, ed. *Recognizing the autonomy of nature: Theory and practice*. Columbia University Press, New York.

Ridder, B. 2007. The naturalness versus wildness debate: Ambiguity, inconsistency, and unattainable objectivity. *Restoration Ecology* 15:8–12.

Rinella, M. J., B. D. Maxwell, P. K. Fay, T. Weaver, and R. L. Sheley. 2009. Control effort exacerbates invasive-species problem. *Ecological Applications* 19:155–162.

Rolston, H. 1999. Nature, spirit, and landscape management. Pp. 17–24 in B. L. Driver, D. Dustin, T. Baltic, G. Elsner, and G. Peterson, eds. *Nature and the human spirit: Toward an expanded land management ethic*. Venture, State College, PA.

Sazenz, D., R. N. Conner, D. C. Rudolph, and R. T. Engstrom. 2001. Is a "hands-off" approach appropriate for red-cockaded woodpecker conservation in twenty-first–century landscapes? *Wildlife Society Bulletin* 29:956–966.

Schroeder, H. W. 1994. Wild metaphors: Nature as machine or person? *The Futurist* March–April:64.

Schroeder, H. W. 2007. Symbolism, experience, and the value of wilderness. *International Journal of Wilderness* 13(1):13–18.

Scott, D. W. 2002. "Untrammeled," "wilderness character," and the challenges of wilderness preservation. *Wild Earth* 11(3/4):72–79.

Seastedt, T. R., R. J. Hobbs, and K. N. Suding. 2008. Management of novel ecosystems: Are novel approaches required? *Frontiers in Ecology and the Environment* 6:547–553.

Sydoriak, C. A., C. D. Allen, and B. F. Jacobs. 2001. Would ecological landscape restoration make the Bandelier Wilderness more or less of a wilderness? *Wild Earth* 10(4):83–90.

Thoreau, H. D. 1862. Walking. *Atlantic Monthly* 9(56, June):657–674.

Turner, J. 1996. *The abstract wild*. The University of Arizona Press, Tucson.

Van Mantgem, P. J., N. L. Stephenson, J. C. Byrne, L. D. Daniels, J. F. Franklin, P. Z. Zule, M. E. Harmon, et al. 2009. Widespread increase of tree mortality rates in the western United States. *Science* 323:521–524.

Watts, R. D., R. W. Compton, J. H. McCammon, C. L. Rich, S. M. Wright,

T. Owens, and D. S. Ouren. 2007. Roadless space of the conterminous United States. *Science* 316:736–738.

Western, D. 2004. Managing the wild: Should stewards be pilots? *Frontiers in Ecology and the Environment* 2:495–496.

Willers, B. 1992. Toward a science of letting things be. *Conservation Biology* 6:605–607.

Willers, B., ed. 1999. *Unmanaged landscapes: Voices for untamed nature*. Island Press, Washington, DC.

Woods, M. 2005. Ecological restoration and the renewal of wildness and freedom. Pp. 170–188 in T. Heyd, ed. *Recognizing the autonomy of nature: Theory and practice*. Columbia University Press, New York.

Zahniser, H. 1956. The need for wilderness areas. *The Living Wilderness* 59(Winter–Spring):37–43.

Zahniser, H. 1963. Editorial: Guardians not gardeners. *The Living Wilderness* 83(Spring–Summer):2.

Chapter 7

Ecological Integrity: A Framework for Ecosystem-Based Management

STEPHEN WOODLEY

> A protected area is a clearly defined geographical space, recognised, dedicated and managed, through legal or other effective means, to achieve the long-term conservation of nature with associated ecosystem services and cultural values.
>
> —*International Union for the Conservation of Nature*

To most people, the spectacular landscape of the Bandelier National Monument and Wilderness in New Mexico still looks largely undisturbed and much as it has for centuries. The plateau and mesa tops are covered with piñon–juniper woodlands and dissected by deep canyons, with volcanic cliffs replete with cliff dwellings. Cactus and wildflowers still bloom among the trees and grasses in spring. As discussed in Chapter 1, however, a more careful examination reveals that all is not well here. A legacy of historic livestock grazing, fire suppression, loss of human and other predators, feral burros, and punctuated periods of drought have had dramatic effects on the landscape. Fire regimes have changed; fire no longer carries across the landscape. This, in turn, has altered such ecological processes as productivity, nutrient cycling, and even the hydrologic cycle. Despite appearances, some important components of Bandelier ecosystems are missing or not functioning properly; the landscape has lost much of its ecological integrity (Sydoriak et al. 2000).

This is a story that is being played out across the landscape of many parks, wildernesses, and protected areas. Banff National Park in Canada has seen dramatic ecological changes, despite persistence of its spectacular scenery and wildlife (Kay et al. 1999; Pyne 2004). A major challenge to Banff's ecological integrity was the large elk (*Cervus canadensis*) population that grew in response to the cessation of aboriginal hunting and a reduction in predation by wolves (*Canis lupus*). Along with fire suppression, hyperabundant elk contributed to the decimation of aspen (*Populus tremuloides*) and willow (*Salix*) communities and populations of species such as beaver (*Castor canadensis*). Using ecological integrity as a management goal, Parks Canada has responded by restoring a mixed human- and lightning-caused fire regime, mimicking aboriginal hunting, and reestablishing predator–prey relations. This is active ecological restoration, guided by a science-based assessment of ecological integrity.

The premise of this book is that new approaches to protected area management are needed in order to conserve nature and biodiversity. Chapter 6 explored autonomous nature as a desirable goal, at least for some protected lands; Chapters 8 and 9 will explore historical fidelity and resilience as desirable goals. This chapter focuses on ecological integrity as a new management goal for protected areas. Although there is some overlap with the goals of historical fidelity and resilience, ecological integrity provides a unique management emphasis, focused on the maintenance of intact, sound, functioning ecosystems. It allows the possibility of active human management of protected areas. It provides the understanding that intentional manipulation of ecosystem processes will be needed to maintain integrity in many protected areas. It acknowledges that the pre-Columbian Americas were populated by millions of aboriginal peoples with cities, roads, and engineering structures (Mann 2005). Even outside the highly populated areas, First Nations and aboriginal peoples were keystone predators, regulating levels of ungulate populations and modifying ecosystems through complex use of fire (Pyne 1983). Unlike the concept of naturalness, ecological integrity assumes that people were, and can be, integral parts of many ecosystems.

This chapter begins with an exploration of how *ecological integrity* is defined. Then I describe how this concept was pioneered in the Canadian national parks. Recognizing some of the limitations of the naturalness concept explored in this book, Canada amended its National Parks Act to make its mission the maintenance of ecological integrity. An emphasis on ecological integrity focuses attention on desirable attributes of protected area ecosystems and what is to be preserved and protected and, importantly, provides more rigorous and measurable management objectives. In this chapter I detail the way ecological integrity is conceptualized, measured, and managed for in Canadian national parks.

What Is Ecological Integrity?

The terms *ecological integrity*, *ecosystem health*, and *biodiversity* have been increasingly adopted by protected area managers to describe the goals for ecosystem management. However, ecological integrity has become most entrenched in the scientific literature, in national and provincial legislation, and in the language of international agreements and treaties. Numerous statutes and official policy statements have articulated ecological integrity as a goal, from the Great Lakes Water Quality Agreement (International Joint Commission 1978) and the Canadian Forest Service to the Millennium Ecosystem Assessment (MA 2006) and the Convention on Biological Diversity (2004). According to the dictionary, to have integrity something must be complete, whole, intact, sound, and unimpaired. When it is qualified by the adjective *ecological*, the implication is that ecological systems must have these attributes. So if ecological integrity is the overriding goal of management, the charge to park managers is to ensure that ecological systems are not missing important parts or structures and that they are functioning well.

Defining *ecological integrity* first requires a definition for *ecosystem*. Conceptually, ecosystems have boundaries that are dynamic and permeable. The species that make up ecosystems respond individually to resource needs and opportunities, with their own spatial variation (Westman 1990). Ecosystems change over short and long time scales with succession and climatic shifts. Application of the ecosystem concept is thus complex, in a world that needs practical solutions. In practical terms, ecosystem managers are concerned with maintaining native biodiversity, intact trophic levels, and the dynamic processes that support native species.

Beyond the definition of an ecosystem, precisely defining ecological integrity can be challenging. There is undoubtedly agreement that a pristine park in Alaska or the Yukon has ecological integrity. It has a full complement of native species, ecological processes and structures are unimpaired, and water and air quality are high. We might also agree that a cornfield in Kansas or Ontario has impaired ecological integrity. As an ecosystem it has lost its species diversity, ecosystem functions are impaired (e.g., rapid nutrient loss), and structures are rapidly degrading (e.g., soil loss). However, beyond the extremes, answers are less clear. The idea requires careful thought and indeed careful measurement.

The notion of ecological integrity has been discussed from many perspectives in collections by Edwards and Regier (1990), Woodley et al. (1993), and Pimentel et al. (2000), and a variety of definitions have been put forth. Kay (1993: 201) concludes that "our sense of what constitutes ecological integrity is very much dependent on our perspective of what constitutes a whole ecological system."

In this early definition, Cairns (1977: 171) included both social and ecological elements and recognized that ecological integrity (or biological integrity) is a function of geographic location: "Biological integrity is the maintenance of the community structure and function characteristic of a particular locale or deemed satisfactory to society." In 1981, Karr and Dudley (1981: 68) defined integrity in more quantitative terms, developing a quantitative index of biological integrity for streams. They defined biological integrity as "the capability of supporting and maintaining a balanced, integrated, adaptive community of organisms having a species composition and functional organization comparable to that of the natural habitat of the region." Noss (1990: 242) stated that "when a community is dominated by native species, is relatively stable and shows other attributes of 'health,' it is said to have integrity."

A definition provided by Kay (1991) focuses on the ability of a system to maintain its organization in the face of changing environmental conditions. He defined integrity in terms of the thermodynamic properties of ecosystems. The successional pathways of ecosystems are equated to a thermodynamic branch. Somewhere along the thermodynamic branch is an optimum operating point, where the thermodynamic properties are optimized. Integrity is lost with deviations from the optimum operating point. The optimum operating point imparts some stability or resistance to change to the system. According to Kay (1991: 491), "if a system is able to maintain its organization in the face of changing environmental conditions, then it is said to have integrity."

More recently, Parks Canada provided the following legal definition of ecological integrity in the 1998 National Parks Act, which tried to consider all these elements. The act states,

Ecological integrity means, with respect to a park, a condition that is determined to be characteristic of its natural region and likely to persist, including abiotic components and the composition and abundance of native species and biological communities, rates of change and supporting processes.

Ecological Integrity in Parks and Protected Areas

Understanding ecological integrity in the context of parks and protected areas requires careful thought about how an ecosystem is structured and how it is functioning. From the science of ecology we understand that ecosystems exhibit a number of important characteristics.

They Should Retain a Full Complement of Native Species

Most protected areas were, and are, established to conserve native species, sometimes expressed as biodiversity. Species loss can result from many kinds of ecosystem stress. Examples of stressed ecosystems losing species include Canadian boreal forests subject to high sulfur dioxide emissions (Freedman and Hutchinson 1980), temperate deciduous forest subject to radiation exposure (Woodwell 1970), and estuarine diatom communities subject to heavy metal pollution (Patrick 1967). In addition to stresses, the other main causes of species loss in protected areas are habitat fragmentation and the protected area simply being too small. For example, western North American parks have experienced extinction rates that are inversely related to park size (Newmark 1995).

Because the majority of existing protected areas are too small to conserve populations of native species, managers must actively manage populations and ecosystems or make the effective size of the protected area and conserved population larger. Active management and intervention in ecosystem processes, where needed to avoid loss of species, are central to sustaining ecological integrity, as are attempts to develop integrated conservation strategies for the regional landscape (Noss 1983; Henderson et al. 1985).

Selected Indicator Species Should Be Viable

Many protected areas do not have good inventories of even the best-known taxa, such as birds and mammals. Even where good inventories exist, they are generally not repeated at regular intervals. A more practical approach is to monitor selected indicator species, or focal species, and track their status. There is a large literature on the selection of indicator species (see Landres et al. 1988; Dufrene and Legendre 1997; Simberloff 1998). Researchers usually determine the status of an indicator species by examining its population vital rates and using those metrics to determine its probability of survival (or, conversely, the probability of extinction), typically for 100 or 1,000 years (Soulé and Simberloff 1986).

The susceptibility of a given species to extinction is a function of many factors; the most important are population size, body size, age at first reproduction, birth interval, and susceptibility to both slow and catastrophic change. Large-bodied, long-lived species such as redwood trees (*Sequoia sempervirens*) have a low rate of turnover and are more susceptible to extinction than small, short-lived species such as annual plants (Goodman 1987). Although there is general agreement on 500 as a minimum viable population size, most researchers agree that this only represents the right order of magnitude; the real minimum might vary between 50 and 5,000 (Lande and Barrowclough 1987).

Ecosystem Trophic Levels Are Intact

Ecosystems have characteristic levels of primary producers, herbivores, and carnivores that can be expressed as food webs. The length of a food web is a characteristic of a specific ecosystem in a specific place. Highly affected ecosystems tend to have food webs that are simple in comparison to those of unmodified ecosystems. In many protected areas, top carnivores such as wolves have been extirpated. This can result in hyperabundant ungulate populations, which have cascading adverse effects on primary producers (White et al. 1998). Another general effect of stress in both aquatic and terrestrial ecosystems is reduction in the average body size of organisms. A decline in body size is generally accompanied by increased prominence of generalist species and a loss of specialist species. Woodwell (1967) found that disturbance resulted in a general trend toward small-bodied, rapidly reproducing, hardy species.

Disturbance Regimes Operate to Maintain Biological Communities with a Mix of Age Classes

Ecosystems are inherently dynamic, driven by fire, climate, weather, and herbivores. After disturbance, ecosystems pass through sometimes predictable successional stages. Repeated disturbance events create a mosaic of biological communities in both time and space. The resulting configuration of community types of different sizes and ages determines the survival of individual species. Thus, the biodiversity of a protected area results from these disturbance factors. Because some disturbances (e.g., fire and herbivory) can be influenced by park managers, this aspect of ecological integrity is under at least partial management control.

Productivity and Decomposition Operate within Limits for System Persistence

Most ecosystems are driven by primary productivity, the amount of organic matter produced by biological activity per unit area in a given time period. Schaeffer et al. (1988) propose that ecosystems be viewed as systems that maintain their structural integrity by degrading energy while avoiding entropy. The onset of ecosystem illness occurs when subtle shifts in productivity occur, and profound disease is indicated when energy is lost from the ecosystem in an uncontrolled manner. For example, pine forests exposed to airborne pollutants invariably experience stunted needle growth and premature loss (Williams 1980; Mann et al. 1980). As production decreases, respiration often increases as energy is diverted to repair. Because ecosystems must recycle nutrients, rate of decomposition is a key controlling factor. In ecosystems stressed by acid precipitation, decomposition decreases (Norten et al. 1980). In other stressed systems, such as clear-cut forests, decomposition rates rise significantly. Productivity and decomposition operate within a range for specific ecosystems. When these vital processes move outside that band, the ecosystem is fundamentally affected and loses its integrity.

Nutrient Cycling Is within Limits for System Persistence

In almost all ecosystems, nutrient availability is a limiting factor, and rates of nutrient cycling are critical to ecosystem function. Ecosystems cycle nutrients and conserve nutrients at characteristic rates. As ecosystems become stressed and lose integrity, they lose their ability to retain nutrients, and they exhibit changes in rates of nutrient cycling and in the relative abundances of nutrient pools (Likens et al. 1978; Weber 1977).

Ecological Integrity and Parks Canada

The original goal for management of Canadian national parks, similar to park management in the United States, was to preserve naturalness. In the last decade, ecological integrity has replaced naturalness as the fundamental endpoint of park management in Canada. The Parks Canada definition of ecological integrity incorporates some elements of historical fidelity (the concept explored in Chapter 8), with its concern for maintaining native species composition and abundance. It also explicitly incorporates resilience (the concept explored in Chapter 9), with its concern for the ability of ecosystems to persist into the future.

The shift from naturalness to ecological integrity recognizes that humans have always influenced and managed many ecosystems, even in protected areas. It shifts attention from cause to effect, from a focus on whether humans caused a particular ecological condition to a concern with desired attributes of future ecological conditions. In fire management, for example, the effects of a fire are considered to be more important than the source of ignition (Lopoukhine 1991). From this perspective, active management does not diminish park goals. Rather, it is necessary for biodiversity conservation, particularly in smaller parks situated in more developed landscapes.

But most critically, the concept of ecological integrity provides a clearer foundation for park management. Protected areas are established for specific reasons, for goals such as the conservation of nature. Consequently, management must have specific objectives. If protected area goals and objectives are not measurable, there is no way of knowing whether management is successful. This is particularly important where active management and intervention in ecosystem processes occur. Ecological integrity provides a framework that allows the translation of broad, often vague nature protection goals into more specific and measurable endpoints, based on desirable ecological conditions. Monitoring and assessment are integral parts of management for ecological integrity. In many ways, the key advantage of management for ecological integrity, rather than "naturalness," is that the former is easier to unambiguously define and measure.

Measuring Ecological Integrity

If protected area managers are to achieve the conservation of nature, whether it is defined as naturalness, ecological integrity, historical fidelity, or resilience, they must be able to measure the outcomes of management efforts. Measurement of outcomes is fundamental to any management process. Yet outcomes have been largely unmeasured for protected areas, perhaps because the outcome of "naturalness" has not proved to be really measurable. A key advantage of using ecological integrity as a management endpoint is that it is scientifically measurable. This section describes how

Parks Canada has approached measuring ecological integrity, but the principles involved are applicable anywhere. The U.S. National Park Service Vital Signs Program is almost identical in approach.

Monitoring implies regular or continuous assessment of one or more variables. It is a set of measures taken in a time series (Croze 1984). Such measures are generally designed to provide information about a system, detect changes, and perhaps manage or control a system. A useful definition of monitoring protected areas has been put forward by Elzinga et al. (1998: 1): Monitoring is "the collection and analysis of repeated observations or measurements to evaluate changes in condition and progress toward meeting a management objective."

Ecological integrity monitoring aims to collect and analyze data on a suite of carefully selected indicators in a rigorous and consistent manner and to compare and report results according to management targets and thresholds. In Parks Canada, each park has selected six to eight major park ecosystems (e.g., forest, tundra, grassland) as indicators. In very practical terms, a small number of indicators works best with parks managers, stakeholders, and the public, who relate to known ecological entities such as forests rather than more esoteric scientific concepts such as productivity. The assessment of an ecological indicator is based on a set of ecological measures, which are the ecological attributes of these major ecosystems. The selection of this suite of measures is based on the following steps:

Step 1. Construct a Conceptual Ecological Model for the Major Park Ecosystems (Indicator)

Conceptual ecosystem models have several important functions. They reduce ecosystem complexity by making explicit the key interactions and flows in the system. Models also acknowledge and relate components of ecosystem biodiversity, processes, and stressors. Finally the models provide an ecological framework for selecting ecological integrity measures, and for developing a scientific assessment of ecological change in the context of the major park ecosystems and ecological integrity indicators (Figure 7.1).

Step 2. Use the Conceptual Models to Select a Set of Ecological Measures

Such measures will provide the necessary diagnosis of the indicator. A suite of ecological measures is selected with the aim of elucidating key elements

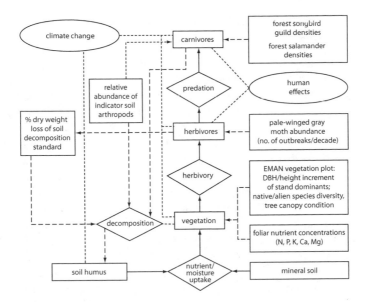

FIGURE 7.1. A conceptual model showing local forest ecosystem (stand-level) components, processes, stressors, and forest measures to be co-located in a replicated series of long-term ecological integrity monitoring plots. DBH = diameter at breast height; EMAN = Ecological Monitoring and Assessment Network.

of the ecosystem, such as structure and ecological function, as well as ecological stressors, at a range of scales. The template used for selection for measures is shown in Table 7.1. The template is based on the ecological attributes that are likely to be responsive to ecological stress. In order to be efficient and effective, only a few measures are selected to support each indicator. Note that this set of measures will never tell a complete story about the state of an ecosystem. They are designed to be a basic diagnostic system. If an ecosystem is losing integrity, additional studies will probably be necessary to determine cause and effect. This is akin to a physician having a basic set of diagnostic tests for general health and a more specific set of tests for a specific disease.

Selecting ecological factors to measure for ecological integrity monitoring is a key decision for program development. Measures should be selected with a long-term view because these measures will be used to inform our assessments of ecological integrity for a very long time. The measures selected must be responsive to stress, information rich, feasible to implement, and able to provide a clear and comprehensive assessment of the evolving state of ecological integrity.

TABLE 7.1. Ecological Integrity Monitoring Template of Measures

Biodiversity	Ecosystem	Stressors
Species Lists	**Succession and Retrogression**	**Land Use Patterns**
Change in species	Disturbance frequencies	Land use maps, road
richness	and size (fire, insects,	densities, popula-
Number and extent	flooding)	tion densities
of exotics	Vegetation age class	**Habitat Fragmentation**
Population Dynamics	distributions	Patch size, inner
Mortality and natality	**Productivity**	patch, distance,
rates of indicator	Remote or by site	forest interior
species	**Decomposition**	**Pollutants**
Immigration and	By site	Sewage, petro-
emigration of in-	**Nutrient Retention**	chemicals
dicator species	Ca, N by site or watershed	Long-range transport
Population viability		of toxins
of indicator species		**Climate**
Trophic Structure		Weather data
Faunal size class		Frequency of extreme
distribution		events
Predation levels		**Other**
		Park-specific issues

Source: Woodley (1993).

Step 3. Validate and Test Measures

All measures undergo an establishment phase to assess their feasibility, cost-effectiveness, and interoperability with other measures. The measures must be feasible in terms of sampling logistics, project costs, and the training needed to conduct the sampling. Finally, how variable are the ecological integrity measures, and what kind of replication will be needed to establish desirable levels of statistical power and significance? Field testing of data collection procedures is a central part of this step. In most protected areas, selected measures will draw from a mix of ongoing monitoring programs, new initiatives, and data from partners or stakeholders.

Step 4. Determine Thresholds for Each Measure

Thresholds represent decision points in an interpretation of the continuous variable of ecological integrity. It is through thresholds that the ecosystem condition is assessed. It is often difficult or impossible to set precise thresholds based solely on scientific evidence (Groffman et al. 2006). In some cases, such as water quality targets or well-studied animal populations, threshold values will already be established. For other ecological integrity measures, there may be a sufficiently detailed long-term dataset for protected areas that can be used to establish sample replication requirements. For many of the new ecological integrity measures it is necessary to establish temporary targets and thresholds based on information from relevant literature or from expert opinion. These targets and thresholds will improve through the accumulation of data and experience over time.

Parks Canada uses thresholds to categorize measures and then indicators into "good," "fair," and "poor" classes, which are used for reporting. For a given indicator (major park ecosystem), a rule set is used to aggregate the results for all supporting measures into a "good," "fair," or "poor" rating.

Steps 5, 6, and 7. Protocol Monitoring, Reporting, and Program Review

After indicators are selected, measures tested, and thresholds developed, specific monitoring protocols are developed. Each park uses monitoring results to produce a "State of the Parks" report every 5 years (Table 7.2). Review and quality control procedures ensure that new knowledge, improved methods, and evolving social and management priorities are integrated into the program.

Managing for Ecological Integrity

The results of ecological integrity assessments are used to make decisions about the kinds of active management and restoration that are needed, if any. In Parks Canada the Park Management Plan is a public accountability document that provides an overall direction for park management. The key actions for ecosystem management are specified here, including active management and restoration. This section presents a few case studies of the kinds of active management and restoration programs used to manage for ecological integrity in Canada's national parks.

Loss of Keystone Predators Leads to the Need to Control Herbivores

One of the key issues facing protected area managers is the loss of keystone predators, such as wolves. In many protected areas this has led to a

TABLE 7.2. Ecological Integrity Summary Report for Gros Morne
National Park

Indicator Ecosystems	Trend	Condition	Percentage of Park Area	Rationale for Rating
Forests	Declining	Poor	44	Extreme moose density is affecting regeneration. Decline in forest connectivity outside the park. High percentage of nonnative mammals. Loss of New-foundland wolf, decline of Newfoundland marten and red crossbill.
Barrens	Declining	Fair	35	Declining woodland caribou, increasing human use.
Wetlands	Declining	Fair	11	Decline of woodland caribou. Snowmobile damage. Increasing nonnative species.
Freshwater	Stable	Good	8.8	Healthy invertebrate populations. Concerns about Atlantic salmon, brook trout, and nonnative rainbow trout.
Seacoast	Improving	Fair	0.2	Only a few pair of terns continue to nest in the park. Dunes and coastal forests are recovering from historic grazing, trampling, and human use.
Marine	Declining	Fair	1	Overexploitation of many species, pollution, garbage.

hyperabundance of herbivores such as whitetail deer (*Odocoileus virginianus*) in Point Pelee, a very small national park in southern Ontario, typical of many protected areas in eastern North America with very abundant white-tailed deer. In the case of Point Pelee, high densities of deer caused undesirable changes to the vegetation, including elimination of several species of rare Carolinian flora and the invasion of many exotic species of plants. Exotic species constitute 60 percent of the entire park flora and dominate many disturbed sites. In 1992, Pelee National Park began a program of reducing deer populations in the park, following a detailed assessment of the situation and public consultation. The reduction is now an ongoing part of park management, with regular culling conducted by park staff. The decision to control deer in Pelee was based on a clearly articulated vision of ecological integrity in the park's management plan. Very high deer populations are inconsistent with protecting ecological integrity, representative of a functioning Carolinian ecosystem.

Restoring Kootenay's Original Dry Grasslands and Open Forests

Kootenay National Park in British Columbia has a dry, low-elevation valley that supports rich biodiversity and critical wildlife habitat. This area's dry Douglas fir, ponderosa pine, and wheatgrass (*Pseudotsuga menziosii, Pinus ponderosa*, and *Agropyron spicatum*) vegetation provides important winter range for wildlife, including the Rocky Mountain bighorn sheep (*Ovis canadensis canadensis*).

The long-term fire regime has been a mix of both lightning and aboriginal ignitions, which maintained a mixture of young, middle-aged, and old forests, shrublands, open meadows, and dry grassy slopes. However, more than a century of fire suppression and loss of aboriginal burning have dramatically changed the area's ecology. Without the regenerative benefits of periodic, low-intensity surface fires, the Columbia Valley has been transformed into an even-aged blanket of mature forest that is encroaching on and dominating the original mosaic of species and habitats. Moreover, the dense forest now overtaking the area sets the stage for catastrophic wildfires, which would alter the ecosystem.

To return ecological integrity to the valley and reduce risk of high-intensity wildfire, Parks Canada is restoring the grassland and open forest biodiversity of the South Kootenays through the Redstreak Restoration Project (also described in Chapter 14). The dramatic first step in restoration is the mechanical harvesting of trees, followed by carefully planned and managed prescribed burns. Ecological integrity provides the framework for such dramatic interventions in a protected area landscape. In previous eras of park management, it would have been unthinkable to have a major mechanical harvest in a national park. Using ecological integrity as a management endpoint, Parks Canada recognized that the ecosystem's fire regime was partially human controlled and managed for that endpoint.

Ecological Integrity Advances Protected Area Management

Ecological integrity advances protected area management by giving a system context to the goal of conserving biodiversity. We know that biodiversity persists in complex ecological systems. The management endpoint, or outcome, for protected areas must reflect the complexity of these systems. Ideally it should be measurable, and it should reflect the best understanding of how these systems function and how they are structured. Ecological integrity aims to incorporate ecosystem science into assessment and management of protected areas.

The ideas of natural, historical fidelity, resilience, and ecological integrity all have elements in common. They also have key differences. As described in Chapter 9, resilience is really part of the basic notion of ecological integrity and part of the definition used by Parks Canada, where an ecosystem with integrity is described as "likely to persist." However, ecological integrity focuses on the structure, function, and composition of an ecosystem. Resilience tends to focus on ecosystem function, with the structural and functional elements being more interchangeable. Thus, I argue that ecological integrity offers a more complete framework for protected area management.

Ecological integrity also has some commonalities with historical fidelity, as described in Chapter 8, but ultimately it is different. Ecological integrity recognizes that protected area management cannot attempt to recreate one particular era because most ecosystems are dynamic and complex. Ecosystem goals must be established while disturbance history, levels of herbivores and predators, and changing regional land use patterns are considered. In some cases it may be possible to restore and maintain an ecosystem similar to one in the past, but it will never be exactly the same. The past is an excellent reference framework but not a complete guidance system. For example, fire may be used to maintain a successional stage, such as a grassland valley bottom that, if left alone, would eventually fill in with trees. However, the open grassland seral stage is best maintained for defined ecological objectives, such as maintaining a specific plant community or providing winter forage, rather than recreating a particular era or point in time.

Ecological integrity also represents a departure from the hands-off approach often implied by naturalness, as described in Chapter 6. It is now well understood that many landscapes in which parks are situated, especially smaller parks situated in developed landscapes, have been highly altered. To the extent that a park may be the last stronghold for a particular species, if it is lost from the park that species could be lost from the larger region too. Therefore, if parks are to include species and ecosystems characteristic of the surrounding natural region, park landscapes and species populations may have to be actively managed in order for species to persist. Active management and restoration, where and when necessary, are a fundamental part of the ecological integrity approach.

To be clear, it is preferable to manage for ecological integrity in large protected areas, where management intervention is not needed. However, in order to compensate for past or current actions, active management is often needed in such areas as fire restoration, species and community res-

---

BOX 7.1. MANAGING FOR ECOLOGICAL INTEGRITY

- Maintaining or restoring the ecological integrity of protected area ecosystems is an expression of the primary goal of protected area management: conservation of nature and its biological diversity. Parks Canada has made ecological integrity—rather than naturalness—their overriding management goal, along with the subsidiary goals of experiencing and learning about park ecosystems.
- Integrity, which means that a system is whole, complete, intact, sound, and unimpaired, implies a concern for protecting all the important parts and the proper functioning of ecosystems.
- Ecological integrity is open to the understanding that humans always have been integral to most of the ecosystems that are conserved by protected areas. Humans were keystone species in most of these ecosystems, and their influence must be accounted for in management.
- Maintaining ecological integrity often entails restoration and active management because most protected area ecosystems are too small or are missing key elements or dynamics.
- Key to managing for ecological integrity is the specification of measurable ecological goals and outcomes. This involves the careful selection of ecological indicators, thresholds of acceptability, and management prescriptions for maintaining integrity.

toration, harvest management, management of hyperabundant native species, or elimination of nonnative species. Active management should occur where there is reason to believe that maintenance or restoration of ecological integrity will be compromised without it.

An ecological integrity approach entails the detailed monitoring of ecological endpoints, which is a conceptual leap forward for protected area management. Like it or not, most park managers are faced with difficult management choices. As a management endpoint, ecological integrity is a significant advance from the notion of "naturalness" in that it forces the use of ecosystem science, in combination with societal wishes, to define and decide on ecosystem goals. The use of ecological integrity as a goal in protected area management recognizes that ecosystems are inherently dynamic and have a history of human intervention and management.

REFERENCES

Cairns, J. 1977. Quantification of biological integrity. Pp. 171–187 in R. K. Ballentine and L. J. Guarraia, eds. *The integrity of water*. U.S. Environmental Protection Agency Office of Water and Hazardous Materials, Washington, DC.

Convention on Biological Diversity. 2004. *Program of work on protected areas*. Retrieved August 2008 from www.cbd.int/protected/.

Croze, H. 1984. Monitoring within and outside protected areas. Pp. 628–633 in J. A. McNeely and K. R. Miller, eds. *National parks, conservation, and development*. Smithsonian Institution Press, Washington, DC.

Dufrene, M., and P. Legendre. 1977. Species assemblages and indicator species: The need for a flexible asymmetrical approach. *Ecological Monographs* 67:345–366.

Edwards, C. J., and H. A. Regier, eds. 1990. *An ecosystem approach to the integrity of the Great Lakes in turbulent times*. Great Lakes Fishery Commission Special Pub. 90-4, Ann Arbor, MI.

Elzinga, C. L., D. W. Salzer, and J. W. Willoughby. 1998. *Measuring and monitoring plant populations*. BLM Technical Reference 1730-1. Bureau of Land Management, Denver, CO.

Freedman, B., and T. C. Hutchinson. 1980. Long term effects of smelter pollution at Sudbury, Ontario on surrounding forest communities. *Canadian Journal of Botany* 58:2123–2140.

Goodman, D. 1987. The demography of chance extinction. Pp. 11–35 in M. E. Soulé, ed. *Viable populations for conservation*. Cambridge University Press, New York.

Groffman, P. M., J. S. Baron, T. Blett, A. J. Gold, I. Goodman, L. H. Gunderson, B. A. Levinson, et al. 2006. Ecological thresholds: The key to successful environmental management or an important concept with no practical application? *Ecosystems* 9:1–13.

Henderson, M. T., G. Merriam, and J. Wegner. 1985. Patchy environments and species survival: Chipmunks in an agricultural mosaic. *Biological Conservation* 31:95–105.

International Joint Commission. 1978. *Revised Great Lakes water quality agreement of 1978*. International Joint Commission, United States and Canada, Ottawa, November 22, 1978·

Karr, J. R., and R. R. Dudley. 1981. Ecological perspective on water quality goals. *Environmental Management* 5:55–68.

Kay, C. E., C. A. White, I. R. Pengelly, and B. Patton. 1999. *Long-term ecosystem states and processes in Banff National Park and the central Canadian Rockies*. Occasional Report 9, National Parks Branch. Parks Canada, Ottawa.

Kay, J. J. 1991. A non-equilibrium thermodynamic framework for discussing ecosystem integrity. *Environmental Management* 15:483–495.

Kay, J. J. 1993. On the nature of ecological integrity: Some closing comments. Pp. 201–212 in S. Woodley, J. Kay, and G. Francis, eds. *Ecological integrity and the management of ecosystems*. St. Lucie Press, Del Ray Beach, FL.

Lande, R., and G. F. Barrowclough. 1987. Effective population size, genetic variation, and their use in population management. Pp. 87–123 in M. E. Soulé, ed. *Viable populations for conservation*. Cambridge University Press, New York.

Landres, P. B., J. Verner, and J. W. Thomas. 1988. Ecological uses of vertebrate indicator species. *Conservation Biology* 2:316–328.

Likens, G. E., F. H. Bormann, R. S. Pierce, and W. A. Reiners. 1978. Recovery of a deforested ecosystem. *Science* 199:492–496.

Lopoukhine, N. 1991. A Canadian view of fire management in the Greater Yellowstone area. Pp. 149–162 in R. B. Keiter and M. S. Boyce, eds. *The Greater Yellowstone Ecosystem: Redefining America's wilderness heritage*. Yale University Press, New Haven, CT.

Mann, C. C. 2005. *1491: New revelations of the Americas before Columbus*. Knopf, New York.

Mann, L. K., S. B. McLaughlin, and D. S. Schriner. 1980. Seasonal physiological responses of white pine under chronic air pollution stress. *Environmental and Experimental Botany* 20:99–105.

Millennium Ecosystem Assessment. 2006. United Nations Environmental Programme. Retrieved August 2008 from www.millenniumassessment.org/documents.

Newmark, W. D. 1995. Extinction of mammal populations in western North American national parks. *Conservation Biology* 9:512–526.

Norten, S. A., D. W. Hansen, and R. J. Compana. 1980. The impact of acid precipitation and heavy metals on soils in relation to forest ecosystems. Pp. 152–164 in *Effects of air pollutants on Mediterranean and temperate forest ecosystems*. General Technical Report PSW-43. Pacific Southwest Forest and Range Experiment Station, Berkeley.

Noss, R. F. 1983. A regional landscape approach to maintain diversity. *BioScience* 33:700–706.

Noss, R. F. 1990. Editorial. *Conservation Biology* 4:241–243.

Patrick, R. 1967. *Diatom communities in estuaries*. American Association for the Advancement of Science Publication 83, pp. 311–315. Springer, New York.

Pimentel, D., L. Westra, and R. Noss, eds. 2000. *Ecological integrity: Integrating environment, conservation, and health*. Island Press, Washington, DC.

Pyne, S. J. 1983. Indian fires: The fire practices of North American Indians transformed large areas from forest to grassland. *Natural History* 92(3):6, 8, 10–11.

Pyne, S. J. 2004. Burning Banff. *Interdisciplinary Studies in Literature and Environment* 11:221–247.

Schaeffer, D. J., E. E. Herricks, and H. W. Kerster. 1988. Ecosystem health: I. Measuring ecosystem health. *Environmental Management* 12:445–455.

Simberloff, D. 1998. Flagships, umbrellas, and keystones: Is single-species management passé in the landscape era? *Biological Conservation* 83:247–257.

Soulé, M. E., and D. Simberloff. 1986. What do genetics and ecology tell us about the design of nature reserves? *Biological Conservation* 35:19–40.

Sydoriak, C. A., C. D. Allen, and B. Jacobs. 2000. Would ecological landscape restoration make the Bandelier Wilderness more or less of a wilderness? Pp. 209–215 in D. N. Cole, S. F. McCool, W. T. Borrie, and J. O'Loughlin, comps. *Wilderness Science in a Time of Change Conference*. Vol. 5: *Wilderness ecosystems, threats and management*. Proceedings RMRS-P-15-VOL-5. USDA Forest Service, Rocky Mountain Research Station, Ogden, UT.

Weber, M. G. 1977. *Nutrient redistribution following fire in tundra and forest–tundra*. M.S. thesis, University of New Brunswick, Fredericton.

Westman, W. E. 1990. Managing for biodiversity. *BioScience* 40:26–33.

White, C. A., C. E. Olmsted, and C. E. Kay. 1998. Aspen, elk, and fire in the Rocky Mountain national parks of North America. *Wildlife Society Bulletin* 26:449–462.

Williams, W. T. 1980. Air pollution disease in the Californian forests: A base line for smog disease on ponderosa and Jeffrey pines in the Sequoia and Los Padres national forests, California. *Environmental Science and Technology* 14:179–182.

Woodley, S. J. 1993. *Assessing and monitoring ecological integrity in parks and protected areas*. Ph.D. thesis, University of Waterloo, Waterloo, ON.

Woodley, S., J. Kay, and G. Francis, eds. 1993. *Ecological integrity and the management of ecosystems*. St. Lucie Press, Del Ray Beach, FL.

Woodwell, G. E. 1970. Effects of pollution on the structure and physiology of ecosystems. *Ambio* 11:143–148.

Woodwell, G. M. 1967. Effects of ionizing radiation on terrestrial ecosystems. *Science* 138:572–577.

Chapter 8

Historical Fidelity: Maintaining Legacy and Connection to Heritage

DAVID N. COLE, ERIC S. HIGGS, AND PETER S. WHITE

History is the witness that testifies to the passing of time; it illumines reality, vitalizes memory, provides guidance in daily life and brings us tidings of antiquity.

— *Cicero*

Telling the future by looking at the past . . . is like driving a car by looking in the rearview mirror.

—Herb Brody

As the story goes, the idea of national parks in the United States emerged one evening in 1870 around a campfire at the junction of the Firehole and Gibbon rivers in Yellowstone National Park (Sellars 1997). Toward the end of an exploratory survey of the region, members of the Washburne–Doane expedition, awed by the spectacles and scenery they had observed, agreed that it would be a travesty if these wonders were despoiled by development and commercialization. Rather, they asserted, the place should be protected and preserved as a public park so all Americans, present and future, could observe the same waterfalls, canyons, geysers, and wildlife and experience the same wild landscapes they had. Within a few years, Yellowstone National Park was created.

This creation myth, cloaked in seemingly altruistic intentions, has often been questioned; utilitarian motives and business interests also played a significant role in the establishment of national parks (Sellars 1997). But there is no doubt that preserving these places much as they are, so their wonders could be experienced by future generations, was a central motivation for the establishment of parks and wilderness areas. John Muir used similar arguments in the Sierra Nevada, as did early proponents of wilderness designation, such as Aldo Leopold and Bob Marshall. This sentiment ultimately was translated into the core mission of the National Park Service, as expressed in its Organic Act of 1916, to conserve scenery, natural and historic objects, and wildlife, leaving them "unimpaired for the enjoyment of future generations."

Originally, humanistic goals such as nostalgia, monumentalism, and the protection of romantic landscapes (Vale 1988) were more fundamental reasons for park and wilderness designation than the ecological goals that have received increasing attention since at least the mid-twentieth century. The park and wilderness movements were largely about passing a legacy of natural and cultural heritage to succeeding generations. Preserving natural conditions in protected areas was widely conceived to be the means for ensuring that future conditions would be faithful to those of the past. Therefore, naturalness often has been equated with the degree to which current conditions are true to the past. As noted in Chapter 2, historical fidelity is one of three primary meanings of naturalness, along with lack of human effect and freedom from intentional human control.

Since the mid-twentieth century, maintaining historical fidelity also has become an important strategy for ecosystem sustainability. Where ecosystems have been degraded, restoration of past conditions is a common objective of interventions designed to conserve biodiversity. As with each management objective discussed in the second part of this book, the pursuit of historical fidelity has promise but also presents challenges. As noted in earlier chapters, the dynamism of ecosystems and the pervasiveness of anthropogenic change make management for historical fidelity more challenging than originally envisioned. Constant change means that maintaining historical fidelity will become increasingly difficult over time. Even more important, if the future is characterized by rapid climate change and no-analog conditions, restoration of past conditions may not be a sustainable option (Harris et al. 2006). Maintaining a high degree of historical fidelity would increasingly be at odds with protecting certain aspects of biodiversity and increasing resilience in protected area ecosystems.

In this chapter, as our opposing epigraphs suggest, we attempt to navigate a narrow divide, both celebrating the value of historical fidelity and expressing concern about its inappropriate use. We define historical fidelity and describe why it is an important management emphasis in protected areas. We explore challenges to its application, particularly ecosystem dynamics, uncertainty about past and future conditions, and the implications of global climate change. We describe how to manage for historical fidelity and articulate ways historical information should and should not be used. We conclude by advocating a balance between the opportunities and challenges of this approach.

Defining Historical Fidelity

The primary definition of the word *fidelity* is "faithfulness to or accuracy in the reproduction of something." We speak of the fidelity of a sound recording when it is a faithful and accurate reproduction of the original performance. So *historical fidelity* implies being true to the past. When protected area managers intervene in ecosystem processes, one of a number of outcomes they might seek is an ecosystem with a high degree of historical fidelity. Such interventions can appropriately be called restorations because they involve restoring conditions to approximations of the past, thereby maintaining legacy, heritage, and authenticity—the historicity of a place (Higgs 2003).

It is common to describe ecosystems in terms of their composition, structure, and function. As Aplet and Keeton (1999) articulate, composition consists of the physical, chemical, and biological objects that make up the ecosystem, including the individual species and their abundances. Structure is the vertical and horizontal distribution of these objects in space, as in the vertical layers of vegetation in a forest or the size and distribution of vegetation patches of differing age. Function consists of the processes that affect ecosystems and through which structure and composition interact. These include both internal processes, such as nutrient cycling and predation, and external processes, many of which are thought of as disturbances, such as fire and flooding.

An ecosystem with high historical fidelity ought to be compositionally, structurally, and functionally similar to past ecosystems found at that place. Each of these ecosystem attributes is important. In comparison to ecological integrity and resilience, composition and structure are more central to historical fidelity. We are not arguing that process is not important. But it is the composition and structure of an ecosystem that typically convey the fidelity of a restoration and give humans an important sense of connection to the past. Moreover, historic fidelity often emphasizes the species that were historically most abundant, iconic, or charismatic, not necessarily the keystone species, the species that are least redundant, or the species most critical to ecosystem function.

When considering historical fidelity as an objective, it is important to view fidelity in a relative rather than absolute sense. Producing a replica of a Stradivarius violin, one of the most highly prized professional musical instruments, is a matter of degree. One approach is to mimic as closely as possible the structure of the instrument by using the same woods, glues, and finishes. Another is to match the feel and sound of the instrument, which may or may not be the same as mimicking the structure of the original. That an artisan may go to great lengths to create a reproduction using similar woods but opt for a slightly different formulation of finish does not mean the work is unfaithful; it means instead that the work approached but did not meet the most rigorous standards for fidelity.

The same is true with ecosystems. Although *historical fidelity* implies an accurate reproduction of the past, no restoration at the spatial scales relevant to protected area ecosystems can be exact. Instead, restorations should be approximations, the precision of which is likely to be variable. What is important is that the natural heritage that comes from the past is carried forward to the present and on into the future. Having said that, there is certainly some point at which the tie to the past is so faint that little fidelity remains.

Reasons for Pursuing Historical Fidelity

There are many reasons for ensuring that the outcomes of ecosystem interventions have some degree of historical fidelity. Some reasons relate to the humanistic goals that were arguably the initial reasons parks and wilderness areas were set aside. Others relate to ecological goals that have emerged more recently and, in at least some places, gained precedence. One of the reasons to sustain past conditions—a reason so obvious as to be almost tautological—is that preservation of park and wilderness ecosystems, their elements and processes, is itself a fundamental goal of management. Certainly no management goal is more fundamental at Joshua Tree National Park or Sequoia National Park, for example, than preservation of the Joshua trees (*Yucca brevifolia*) and giant sequoias (*Sequoiadendron giganteum*) for which the parks were established. But beyond the icons, the lesser species of flora, the fauna, the abiotic environment, and the processes that link and sustain them are all of import. They are interlinked and interdependent. An emphasis on historical fidelity is a means of keeping all the parts of well-functioning ecosystems.

Cultural Reasons to Maintain or Restore Historical Fidelity

There are many significant cultural reasons for attending to history in order to understand ecosystems, and these qualities amplify the ecological value of historical knowledge. We propose that these reasons fall into three categories: nostalgia, place, and time depth. Nostalgia is a bittersweet longing for the past (Higgs 2003). Such longing is never unvarnished in the sense that the past is conclusively better. Memory selectively recalls the qualities that were favorable and pushes the negative aside. Scratch the surface and you will see that returning to one's fond childhood experiences, for example, might be underlain by family conflict that appears only later. As much as anything, this represents a quest for simpler lives in less hurried times. The naiveté in such longing is what lends nostalgia a superficial connotation.

When it comes to nostalgia for ecosystems, the appeal of past systems is that they provide models of nature that are more consistent with affection for wildness: standing on the Great Plains in the 1860s watching a horizon-to-horizon herd of bison, seeing the sky darken with flights of the now-extinct passenger pigeon, walking down the final slope to the Pacific Ocean in northern California through untrammeled old-growth forests. These and other unattainable experiences animate the desire to intervene in ways that create approximations of the conditions.

Place is a second reason why historical fidelity is important. A sense of place is created through the telling of stories, whether lyrical or scientific. Through stories, personal and collective meaning is attached to particular locations. At an individual level, one can feel this connection through the patterns of daily activities—a special trail along which one walks or the location of a series of long-term ecological experiments. At a collective level, people vest certain locations with deep significance. National parks are such places. They are remarkable in their own right, but they are given grace through activities and stories, and over time they become more than just features; they emerge as places. The fabled geysers of Yellowstone are perhaps most significant in this respect. As icons of wilderness, the geysers have become cultural reference points. Surely as they are ecological and geophysical features, so too are they now cultural ones.

Places acquire significance because of their connection to the past, as when one returns to the places of childhood to reinforce family memories or recollect events that helped shape personal identity. A sense of place does not require direct personal connection to a particular location. As Higgs (2003: 152) suggests, "Once we understand that place matters, in other words that we have found within our own lives the qualities that make a place, it is easier to regard these qualities in other places." The act of intervening in ecosystems is motivated in part by the desire to create connections that nurture a sense of place, and the conservation and preservation of ecosystems are similarly motivated by a desire to maintain places.

Finally, ecosystems perceived to have remained the same for long periods of time, which have not experienced recent disruption or human simplification, have time depth. An old-growth forest, for example, or an untilled grassland ecosystem possesses a rarity born of profound historical continuity, providing humans with an opportunity to glimpse a past that seems otherwise mysterious and unknowable. Climate change threatens such continuity by pushing ecosystems outside the range of historical conditions. If new conditions develop slowly, it may be possible to incorporate such gradual change into our collective idea of continuity and rarity. However, should the species assembly fundamentally change, and quickly, then our appreciation of these ecosystems may diminish.

Together, nostalgia, place, and time depth constitute historicity, or the condition of being historical (Higgs 2003). Historicity is important because people are joined to ecosystems through emotional connections, cultural ties, and moral values. A final human value of pursuing historical fidelity is the constraint that comes from gathering information and considering the past. The pressure of rapid environmental change generates swift responses. Operating at a vast scale under tight timelines, interventions will need some resistance. One means for providing necessary constraint is the exercise of gathering historical information. The practice of understanding the trajectories of an ecosystem increases respect for the complexity of ecosystems and allows a pause before we dig in.

Ecological Reasons to Maintain or Restore Historical Fidelity

Maintaining historical fidelity may also be important to the protection of critical ecological values. A fundamental premise of conservation biology is that over evolutionary time periods organisms have adapted to landscape conditions and disturbance events of the past. This suggests that the potential for survival of these organisms will be reduced if future environmental conditions deviate too much from those of the past. Consequently, restoration of past conditions has often been advanced as a means of ensuring sustainability. For example, Manley et al. (1995: xiii) assert that "restoring and maintaining landscape conditions within distributions that organisms have adapted to over evolutionary time is the management approach most likely to produce sustainable ecosystems." According to this theory, then, managing for historical fidelity is a strategy for protecting biotic diversity and the resilience of protected area ecosystems. This has led managers to restore the structure of forested ecosystems, where the absence of fire converted park-like forests to dense tangles of undergrowth (Figure 8.1).

The scientific basis for these assumptions comes from various fields, particularly from disturbance ecology. Karr and Freemark (1985) argue

FIGURE 8.1. Fire and other interventions can be used to restore the widely spaced, open-grown structure of pine forests that were common before the era of fire suppression. This forest near Miquel Meadow in Yosemite National Park was prescribed burned in 1976 to restore historic conditions and reburned by wildfire in 1996. (Photo taken in 2002 by Jan van Wagtendonk)

that ensuring that disturbance regimes continue to operate as they have in the past is crucial to the preservation of genetic, population, and assemblage dynamics. Numerous studies have documented loss of species and adverse ecosystem changes in places where "natural" disturbance regimes and habitats have been substantially altered (Swanson et al. 1994). Landres et al. (1999: 1180) observe that "contemporary anthropogenic change may diminish the viability of many species adapted to past or historical conditions and processes," that "approximating historical conditions provides a coarse-filter management strategy that is likely to sustain the viability of diverse species, even those for which we know little about," and that "because of limited understanding about ecosystems, approximating past conditions offers one of the best means for predicting and reducing impacts to present-day ecosystems."

Recently, however, for reasons explored in Chapters 3 and 4, many of these assumptions and premises have been questioned (Millar and Brubaker 2006). Although we remain convinced that there is a place for historical fidelity in protected area stewardship, it is important to understand the implications of doing so—to consider the ecological appropriateness and feasibility of managing for historical fidelity.

Challenges to Managing for Historical Fidelity

One concern often voiced about historical fidelity is that recreating the past ignores the inherent dynamism of ecosystems. The assumption, originally held by many ecologists and still held by many laypersons, that nature is stable, static, and in balance has been replaced over the past half century by a view of nature in flux (Pickett et al. 1992). Protected area policies and perspectives on historical fidelity are slowly evolving in response to this shift.

When the influential Leopold report (1963), reviewing management policies of the National Park Service, recommended that the goal of interventions should be to recreate "the ecologic scene as viewed by the first European visitors," it caused confusion. The report seemingly called for maintenance of static conditions, freezing environmental conditions at a previous stage. When asked to clarify his intentions, however, Leopold stated that it was critically important to allow the free play of dynamic ecosystem processes (Rydell 1998). Today, National Park Service Management Policies (NPS 2006: 4.1) recognize that "natural change" is an "integral part of the functioning of natural ecosystems" and that "natural processes and species are evolving, and the Service will allow this evolution to continue."

The concept of historical range of variability was developed to recognize the importance of disturbance and to account for and accommodate the inherent dynamics of ecosystems (Egan and Howell 2001). Although it is sometimes used as a metric and used prescriptively, we are most comfortable using the concept of historical range of variability as a descriptor of both the magnitude of variability in ecosystem properties over time and the historical bounds to those properties. The idea is that ecosystems have always changed over time, responding to disturbance, but always within limits. A dynamic view of recreating the past demands that management allow for change in response to disturbance while ensuring that conditions are not pushed beyond the bounds defined by the historical range of variability. In a Rocky Mountain forest, for example, certain changes in species, abundances, and densities would be acceptable; other shifts would not be. As Pickett et al. (1992: 82) conclude, "nature has functional, historical, and evolutionary limits. Nature has a range of ways to be, but there is a limit to those ways, and therefore human changes must be within those limits."

A related idea is that we can more appropriately incorporate ecosystem dynamism into restoration by understanding and restoring ecological trajectories. When degradation occurs, ecosystem conditions change, which in turn alters the present and future trajectories of the system. Rather than recreate the specific conditions that were present at a particular time, an alternative management goal might be to intervene so that trajectories become more consistent with those of the past. This bears some similarities to what Millar et al. (2007) call realignment.

Although some concerns about historical fidelity can be alleviated if we adopt a relative rather than absolute perspective on the term, advance the concept of historical range of variability, and restore ecological trajectories rather than specific conditions, other concerns cannot. Problems with the feasibility of managing for historical fidelity remain. Given the specter of global change, maintaining and restoring historical fidelity will increasingly be a Sisyphean task. Restorations are likely to entail substantial ongoing management intervention to keep ecosystems from adjusting to the new conditions of the future. Costs may be prohibitive. Ecosystems will continue to be assaulted by such stressors as invasive species and pollution and may have lost key natural disturbance processes or even keystone species. As noted in the next section, data needed to understand past ecosystems may be missing or misleading.

Of at least as much concern is that, as explored in Chapter 4, the directional nature of climate change suggests that the abiotic conditions of the future may have no analog. If so, efforts to sustain historical fidelity might reduce the sustainability and resilience of ecosystems, jeopardizing the values managers seek to protect and preserve. All this suggests that although historical fidelity is an important protected area value, managing to maintain or restore past conditions will not be possible everywhere. Moreover, even where it is possible to maintain historical fidelity, doing so will often not be desirable or appropriate.

Using Historical Data in Ecological Restorations

Historical information about past ecosystems is important to ecological restoration, regardless of the degree of historical fidelity that is desired. Data on past conditions and processes reveal the key underpinnings of ecological systems, suggest appropriate trajectories, and provide insight into important ecosystem drivers. They provide one of several important sources of information that can be used to develop an understanding of reference conditions (White and Walker 1997). Among other things, reference conditions are used, with varying degrees of precision, to set restoration goals and targets, providing guidance for interventions and measures of success. It is for this latter purpose that concerns have been raised about using historical data too prescriptively and precisely, particularly given the implications of rapid climate change (Landres et al. 1999; Millar and Woolfenden 1999).

A wide variety of approaches to collecting, interpreting, and applying historical ecological data exist. Egan and Howell (2001: 15) differentiate between "culturally-derived evidence, such as documents, maps, photographs, oral history and Native American land management practices" and "biological (earth-relic) records, including standing woodlots, tree rings, pollen, packrat middens, opal phytoliths, animal remains, and records of changes in soil and hydrology." Given these diverse sources of information, Egan and Howell (2001: 14) recommend that practitioners use a "multiscale, multisource, cross-referential historical analysis that is compared to contemporary data to set reference conditions" and remain open to new information and flexible regarding use of data.

Problems with Historical Data

Two problems with historical data are incompleteness of information about the past and the time and space scale dependence of reference information (Egan and Howell 2001). Incomplete information is problematic for many reasons, particularly because it can lead to interventions that restore a high degree of historical fidelity for ecosystem attributes for which there is substantial historical data but low fidelity for other attributes. The fact that historical data on the structure of tree species and characteristics of fire processes are more abundant than historical data on understory vegetation characteristics and nutrient cycling processes may partially explain why stand structure and fire have received more attention in restoration than other attributes. One should not assume that precise replication of certain ecosystem attributes, such as stand structure or fire frequency, will result in restorations with a high degree of overall fidelity.

As White and Walker (1997: 338) note, "all reference information is inherently time- and space-based." Any characterization of historical range of variability will be valid only for a selected time period and geographic extent (Landres et al. 1999). Narrowing the period of time and the spatial scale to be considered adds precision but introduces other concerns. Resultant conclusions will vary greatly depending on the narrow period of time considered relevant, a decision that often is highly arbitrary. At the other extreme, if very large temporal and spatial scales are used, the range of variation is so great that almost any condition is within appropriate bounds.

Appropriate Use of Historical Data

To address these problems, it is helpful to use as many sources of information as possible and not be overly precise in the use of historical data (Millar and Woolfenden 1999). In other words, it is helpful to use historical data more as descriptive information that provides insight into possibilities and limitations rather than as prescriptive targets or articulations of what ought to be. Although such admonitions work for interventions that are not seeking a high degree of historical fidelity, they are less useful where a high degree of fidelity is desired. In such situations, more useful cautions might be that such interventions are likely to entail perpetual manipulation and that high historical fidelity probably comes at some cost to ecological integrity and resilience.

Despite admonitions about the prescriptive use of historical data, the reality of global change elevates the importance of historical data as a means of informing management response to change. As Swetnam et al. (1999: 1201) observe, knowing history is important "because it informs us about what is possible within the context of certain locations and times, and it places current conditions into this context," as well as providing information "about the potential causes of change and the historical pathways that brought ecosystems to their current condition." Historical data provide information about past conditions, their variability, and how ecological systems have responded to change. A good example is recent work documenting how small mammals have responded to a century of climate change across a 3,000-meter elevation gradient in Yosemite National Park. Moritz et al. (2008) report that half of twenty-eight species monitored showed substantial upward changes in elevation limits. Ranges typically expanded upward for low-elevation species and contracted upward for high-elevation species, although individual species responded idiosyncratically. The research identified several species of concern but concluded that species diversity has changed little, thanks to effective protection of a large-scale elevational gradient.

Historical data are critical in understanding drivers and mechanistic controls of ecosystem processes. For example, such data can resolve divergent interpretations regarding whether fire regimes are driven by seasonal to decadal climatic variations or by local ecological phenomena, such as patterns of vegetation and fuel. A paleoecological perspective indicates that both are important controls of fire regimes (Gavin et al. 2007). Historical data contribute to a better understanding of feedbacks between changes in land cover and regional climate and to improvement of ecological models (Willard and Cronin 2007). Such data can contribute to resolution of stewardship issues such as biological invasions by providing a longer-term perspective on the distinction between what is native and what is not (Willis and Birks 2006). Finally, they can provide insights into thorny questions regarding ecological thresholds, points beyond which management intervention is needed to avoid abrupt, perhaps catastrophic changes in ecosystem quality.

Incorporating Historical Fidelity as a Goal in Ecological Interventions

Arguably the original goal of protected area stewardship, maintenance of historical fidelity remains an important goal today, despite challenges posed by ecosystem dynamism and the pace of directional climatic and other environmental changes. These challenges suggest the need to view historical fidelity in a relative rather than absolute sense, seldom attempting a precise restoration of past conditions. They also suggest the need to emphasize historical fidelity more in certain situations than in others.

Maintaining a high degree of historical fidelity is akin to paddling upstream into the strong current of climate change. In the context of fire restoration, Limerick (2008: 45–46) expresses this challenge as follows:

> Wildland fire presents a prime opportunity to recalibrate the setting on humility and confidence. What are our actual powers? Where does confidence cross into hubris? And if humility needed more to recommend it, global climate change presents precisely that recommendation. Even as we explore the prospect of restoring natural fire to the West's forests, a changing climate makes it impossible to recapture the circumstances of the past, leaving the concept of restoration floating free of an identifiable baseline, an original state to recapture and recreate. If there is anything left to the notion of the forests as a great laboratory, then climate change has made them a laboratory where someone keeps fiddling with the thermostat.

As will be discussed further in Chapter 9, maintaining historical fidelity may require substantial resources and ongoing effort and may reduce ecosystem resilience. Consequently, attempts to emphasize historical fidelity are most likely to be successful where the geographic size of the proposed intervention is small. A historically accurate restoration of a single campsite, for example, is more likely to be successful than restoration of thousands of hectares of pinon–juniper woodland. However, even for small-scale restorations, much of the apparent fidelity may be an illusion if landscape linkages and context are missing where degraded lands surround the restoration. Interventions that emphasize historical fidelity are also more likely to be successful where the objects being restored are less sensitive to change or have been less strongly influenced by directional change. Finally, the importance of emphasizing historical fidelity increases as the rarity and value of the objects of restoration increase. For example, historical fidelity is likely to be more of an emphasis where Joshua tree woodlands and sequoia forests have been degraded than in adjacent communities of less iconic value. Such remnant ecosystems are so redolent with cultural values that heroic interventions will probably be used to maintain historical processes or trajectories.

One example of an intervention that emphasizes historical fidelity is the restoration of 25 hectares of giant sequoia–mixed conifer forest in the Giant Forest Grove of Sequoia National Park. Visitor facilities, including a gas station, a market, hundreds of cabins, campgrounds, and a sewage treatment plant have been removed (Demetry 1998). Because these developments created gaps in the forest, restoration planners decided to use fire-caused gaps as a model for desired restoration outcomes. Specifically, they attempted to restore forest composition and structure such that they mimicked the effects of fire 10 years later. Scientists collected data on the current characteristics of fire-caused gaps in giant sequoia–mixed conifer forest, particularly the horizontal and vertical distribution of woody plant species (Demetry 1998). Based on these data, a prescription was developed for each development-caused gap. Prescriptions specified species composition, density, and spatial patterns within the range of variability found in fire-caused gaps. Treatments varying in intensity were experimentally applied to identify the minimal level of intervention needed to approximate the reference conditions (Demetry 1998). As a coda to this example, illustrative of how perspectives on historical fidelity have changed in the face of climate change, some proponents of this restoration effort now wonder whether it would have been better to use lower-elevation species rather than the species that lived there in the past (Nate Stephenson, personal communication, 2009).

Historical Fidelity and Other Stewardship Goals

We began this chapter celebrating historical fidelity as one of the most traditionally important purposes and values for parks and protected areas. We countered this notion by exploring potential problems with maintaining historical fidelity. We noted that historical fidelity has traditionally been promoted as a strategy for ecosystem sustainability but then voiced concern that, in the face of climate change, restoration of past conditions might lead to loss of resilience. These seemingly contradictory passages might leave the reader wondering where we stand on historical fidelity. We believe that historical fidelity is an important management goal in some places and that adverse outcomes are likely if history is ignored. However, given the pace of global change, interventions that seek a high degree of historical fidelity should be more the exception than the rule in parks and wilderness.

A central thesis of this book is that a diversity of management goals and strategies, including autonomous nature, historical fidelity, ecological integrity, and resilience, must be applied in different parts of individual protected areas and across a protected area system. Planned diversity will optimize the aggregate value of parks and wilderness. It will spread risk and hedge bets in the context of uncertainty and novel stressors, most notably climate change. Thus, we imagine that some interventions will emphasize historical fidelity, as is the case in the example from Sequoia National Park.

Diversity could be enhanced further if we manage some places such that historical fidelity, ecological integrity, resilience, and perhaps even autonomous nature are compromised to equal degrees. That is, the desired outcome in these places would be a modest degree of each management emphasis. Perhaps certain elements of past ecosystems would be sustained while others would be abandoned, if doing so was deemed to enhance long-term resilience. Complementarity with other approaches might also be furthered by the use of historical fidelity to constrain the other stewardship approaches explored in this book as well as to increase their success. For example, historical data are likely to provide insights about ecosystem function that make efforts to promote ecological integrity more successful, and interventions to promote resilience are likely to be more appropriate if they are built around elements that are as true to history as possible.

BOX 8.1. MANAGING FOR HISTORICAL FIDELITY

- Restoring historical fidelity to ecosystems degraded by anthropogenic change is one of several goals or management emphases for ecosystem interventions.
- An emphasis on historical fidelity implies a faithful restoration of past conditions. Although historical processes are important, fidelity to past composition and structure is essential (and thus differentiates this approach somewhat from ecological integrity and resilience).
- Historical fidelity is critical to certain humanistic park and wilderness purposes, particularly nostalgic connections to place and the past, and can contribute to the conservation of biotic diversity.
- Attempts to maintain past conditions are constrained by both the inherent dynamism of ecosystems and the pace of recent directional climatic change.
- Historical ecological data are better treated as being informative rather than prescriptive.
- Historical fidelity should be viewed as relative rather than absolute. Higher degrees of fidelity are appropriate where interventions are small, rare and valued objects of preservation are at risk, and the magnitude of unavoidable directional change is small.
- In addition to being a goal in its own, a concern for historical fidelity can constrain the specific actions taken to promote ecological integrity and resilience (e.g., by giving preference to historic rather than novel elements of biodiversity).

REFERENCES

Aplet, G. H., and W. S. Keeton. 1999. Application of historical range of variability concepts to biodiversity conservation. Pp. 71–86 in R. K. Baydack, H. Campa, and J. B. Haufler, eds. *Practical approaches to the conservation of biological diversity.* Island Press, Washington, DC.

Demetry, A. 1998. A natural disturbance model for the restoration of Giant Forest Village, Sequoia National Park. Pp. 142–159 in W. R. Keammerer and E. F. Redente, eds. *Proceedings of High Altitude Revegetation Workshop no. 13.* Information Series no. 89. Colorado Water Resources Research Institute, Fort Collins.

Egan, D., and E. A. Howell. 2001. *The historical ecology handbook: A restorationist's guide to reference ecosystems.* Island Press, Washington, DC.

Gavin, D. G., D. J. Hallett, F. S. Hu, K. P. Lertzman, S. J. Prichard, K. J. Brown, J. A. Tynch, P. Bartlein, and D. L. Peterson. 2007. Forest fire and climate change in western North America: Insights from sediment charcoal records. *Frontiers in Ecology and the Environment* 5:499–506.

Harris, J. A., R. J. Hobbs, E. Higgs, and J. Aronson. 2006. Ecological restoration and global climate change. *Restoration Ecology* 14:170–176.

Higgs, E. 2003. *Nature by design: People, natural process, and ecological restoration.* The MIT Press, Cambridge, MA.

Karr, J. R., and K. E. Freemark. 1985. Disturbance and vertebrates: An integrative perspective. Pp. 153–168 in S. T. A. Pickett and P. S. White, eds. *The ecology of natural disturbance and patch dynamics.* Academic Press, San Diego, CA.

Landres, P. B., P. Morgan, and F. J. Swanson. 1999. Overview of the use of natural variability concepts in managing ecological systems. *Ecological Applications* 9:1179–1188.

Leopold, A. S., S. A. Cain, D. M. Cottam, I. N. Gabrielson, and T. L. Kimball. 1963. Wildlife management in the national parks. *Transactions of the North American Wildlife and Natural Resources Conference* 28:28–45.

Limerick, P. N. 2008. Fire alarm: Historians, and Thorstein Veblen, to the rescue. *Forest History Today* Fall:40–46.

Manley, P., G. E. Brogan, C. Cook, M. E. Flores, D. G. Fullmer, S. Husari, T. M. Jimerson, et al. 1995. *Sustaining ecosystems: A conceptual framework.* USDA Forest Service, San Francisco, CA.

Millar, C. I., and L. B. Brubaker. 2006. Climate change and paleoecology: New contexts for restoration ecology. Pp. 315–340 in D. Falk, M. Palmer, and J. Zedler, eds. *Foundations of restoration ecology: The science and practice of ecological restoration.* Island Press, Washington, DC.

Millar, C. I., N. L. Stephenson, and S. L. Stephens. 2007. Climate change and forests of the future: Managing in the face of uncertainty. *Ecological Applications* 17:2145–2151.

Millar, C. I., and W. B. Woolfenden. 1999. The role of climate change in interpreting historical variability. *Ecological Applications* 9:1207–1216.

Moritz, C., J. L. Patton, C. J. Conroy, J. L. Parra, G. C. White, and S. R. Beissinger. 2008. Impact of a century of climate change on small-mammal communities in Yosemite National Park, USA. *Science* 322:261–264.

National Park Service. 2006. *Management policies 2006*. Retrieved October 11, 2008 from www.nps.gov/policy/mp/policies.html.

Pickett, S. T. A., V. T. Parker, and P. L. Fiedler. 1992. The new paradigm in ecology: Implications for conservation biology above the species level. Pp. 65–88 in P. L. Fieldler and S. K. Jain, eds. *Conservation biology: The theory and practice of nature conservation, preservation, and management*. Chapman and Hall, New York.

Rydell, K. L. 1998. A public face for science: A. Starker Leopold and the Leopold report. *The George Wright Forum* 15(4):50–63.

Sellars, R. W. 1997. *Preserving nature in the national parks: A history*. Yale University Press, New Haven, CT.

Swanson, F. J., J. A. Jones, D. O. Wallin, and J. H. Cissel. 1994. Natural variability: Implications for ecosystem management. Pp. 80–94 in M. E. Jensen and P. S. Bourgeron, coords. *Ecosystem management: Principles and applications*. Vol. II: *Eastside forest ecosystem health assessment*. General technical report PNW-GTR-318. Pacific Northwest Research Station, Portland, OR.

Swetnam, T. W., C. D. Allen, and J. L. Betancourt. 1999. Applied historical ecology: Using the past to manage for the future. *Ecological Applications* 9:1189–1206.

Vale, T. R. 1988. No romantic landscapes for our national parks? *Natural Areas Journal* 8:115–117.

White, P. S., and J. L. Walker. 1997. Approximating nature's variation: Selecting and using reference information in restoration ecology. *Restoration Ecology* 5:338–349.

Willard, D. A., and T. M. Cronin. 2007. Paleoecology and ecosystem restoration: Case studies from Chesapeake Bay and the Florida Everglades. *Frontiers in Ecology and the Environment* 5:491–498.

Willis, K. J., and H. J. B. Birks. 2006. What is natural? The need for a long-term perspective in biodiversity conservation. *Science* 314:1261–1265.

Chapter 9

Resilience Frameworks: Enhancing the Capacity to Adapt to Change

ERIKA S. ZAVALETA AND F. STUART CHAPIN III

> It's not the strongest of the species that survives, nor the most intelligent that survives. It is the one that is the most adaptable to change.
>
> —*Charles Darwin*

How can we sustain the basic functioning and character of parks and wilderness areas over the very long term? Managing for resilience calls for a shift from pursuit of specific landscape or ecological objectives toward the core goal of sustaining functional, adaptable systems that can deliver core, desired values such as native biodiversity. From this perspective, resilience is a valuable guiding principle in a time of rapid and unpredictable environmental change. Regardless of what other specific goals are managed for, a critical consideration is whether protected areas will be able to weather unprecedented rates and types of change.

As described in more detail in Chapter 4, the earth today is experiencing unprecedented rates and types of directional stresses in both environmental and social realms (Vitousek et al. 1997; Millennium Ecosystem Assessment 2005). Protected area ecosystems must somehow absorb an onslaught of climatic, atmospheric, and biotic changes (such as exotic invasive species) if they are to persist. They must do this in the context of rapidly changing recreational use, fragmentation and development of adjacent landscapes, and eroding societal engagement with nature (Pergams and Zaradic 2008). Past management has not focused on building resilience and has at times and in places applied an engineering paradigm to ecosystem management that actually erodes resilience by suppressing dynamism and disturbances (Holling 1973).

Protected areas today are largely isolated in the landscape, cut off from larger-scale migration, missing a full range of natural disturbances and native biodiversity, and thereby limited in their ability to support adaptive evolutionary and ecological changes. To varying degrees, they have lost resilience. How many of these protected areas will break down without efforts focused on bolstering their ability to adapt? From increasingly fire-prone forest ecosystems to steadily eutrophied wetlands and lakes to meadows and grasslands experiencing successive waves of exotic invasion, protected areas need to be managed with resilience in mind. In this chapter, we define resilience, describe the key features of resilient ecosystems, and explore the ways in which resilience can be integrated into protected area management.

Resilience and Resilient Systems

The concept of resilience has long been applied in fields ranging from engineering to psychology, with a range of meanings. Previous conceptions of resilience focused on the rate at which a system could return to a particular equilibrium after disturbance (Holling 1973). Canadian ecologist C. S. "Buzz" Holling (1973) first distinguished the concept of ecological resilience by emphasizing the capacity of a system to weather shocks and disturbances without shifting or collapsing into a qualitatively different state (Figure 9.1). Holling's perspective reflected a shift from a static to a dynamic perception of ecosystems, as discussed in Chapter 3. From this perspective, complex ecological systems, affected by both an internal web of relations and external variability and change, differ entirely from engineered systems marked by constancy, controllability, and predictability. A resilient ecosystem may vary widely through time, but it retains core functions and the characteristic relationships that sustain them.

A system's resilience has also been defined in terms of three distinct features: the amount of change it can experience while retaining the same controls on its structure and function, the degree to which it is capable of self-organization, and its capacity for learning and adaptation (Resilience Alliance 2008). The first of these allows the system itself to change greatly.

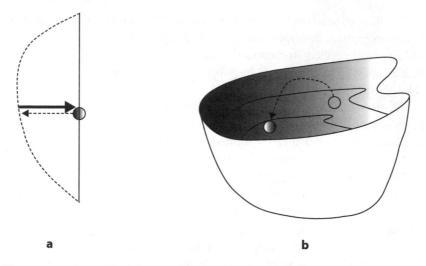

a

b

FIGURE 9.1. (a) Engineering resilience and (b) ecological resilience. In (a), management strives to maintain a constant steady state and to maximize efficiency by emphasizing how quickly an ecosystem (the ball) can return to that steady state after a disturbance (dashed arrow). In (b), management strives to keep the system (the ball) anywhere within broadly defined bounds of characteristic functioning (the cup). In (a), resilience is the rate at which the system returns to its equilibrium. In (b), resilience is the size, shape, and steepness of the cup—its ability to hold the moving ball.

A montane landscape might experience widely ranging locations and relative abundances of forest, meadow, and shrubland patches of different ages and compositions in response to climate, fire, and flood cycles; its resilience is related to the ability of those drivers to return it to a previous configuration. The second of these relates to the ability to recover functioning. After a hurricane in a tropical forest, can vegetation recolonize remaining soils and rebuild them? The third, in an ecological context, includes the capacity for evolutionary responses in populations and ecological responses such as shifts in species distributions or forest composition. Although this chapter's focus is ecological, all of these features apply equally to the resilience of social components of a linked system such as a national park or refuge. The capacity to learn and adapt as an agency, to reorganize into an effective institution after a major policy change, and to weather shifts in political power without abandoning fundamental agency goals and purpose are all marks of institutional resilience that can be as important to the long-term sustenance of protected areas as ecological resilience.

Resilience as a Means and an End

A core question to answer when managing for resilience is, "Resilience of what, to what?" (Carpenter et al. 2001). Undesired systems or states can be extremely resilient, such as the cheatgrass-invaded Great Basin (Mack 1981) or (for several decades, at least) the former Soviet Union (Walker and Salt 2006). Thus, resilience is not inherently good. And people do not visit parks and wilderness to experience resilience. They seek out values that managers strive to make resilient, such as native biodiversity and large, un-transformed landscapes. In the context of protected area management, one must define what should be resilient—what core, desired attributes or functions define the system to be maintained. Resilience can be a means to sustaining values such as ecosystem services, native biodiversity, and aesthetic landscapes. It can also be an end in itself, as in the pursuit of more resilient parks, as long as it is with reference to desired values or attributes.

Managing for truly long-term resilience might call for letting go of the way landscapes look today. As conditions change, for instance in climate, one has to ask what functions or processes one really wants to retain in the face of change so that all the other variables can shift around to maintain the resilience of those core processes. For example, rapidly changing species distributions might shift a goal of maintaining giant sequoia (*Sequoiadendron giganteum*) forest cover to one of maintaining native vegetative cover sufficient to provide diverse habitat for wildlife, maintain soils, and maintain endogenous disturbance regimes. It is also important to remember that resilience does not refer to the speed with which a system returns to being within desired bounds. Under climate change, maintaining the resilience of a protected area could entail allowing the defining species in the park to migrate entirely outside its boundaries for centuries. One would then need to ensure the persistence of those species somewhere in the regional biodiversity pool, to retain the option of having those species recolonize the park hundreds or thousands of years later.

Resilient Ecosystems

We turn now to a detailed example, to elucidate the characteristics of a resilient ecosystem. California's Sierra Nevada montane forests historically experienced wildfires at the scale of landscape patches, creating a mosaic of stands at different stages in a recurring cycle of fire, reseeding, regeneration, fuel buildup, and more fire (Sierra Nevada Ecosystem Project 1996). This mosaic is superimposed on and shaped by a mosaic of different tree species, soil conditions, human activity, and settlement and roads that in turn define varying degrees of fire risk across the landscape. Patches are distinct but also linked by many processes, including the possibility of fire spread across them. Over decades, fire policy and global trends have shaped the character of Sierran forests. Fire suppression policies have increased the similarity of forest patches over larger and larger areas by erasing the heterogeneity once maintained by frequent, patchy fires (Sierra Nevada Ecosystem Project 1996). Climate change has produced longer and more dangerous fire weather each summer, exacerbating the risk of severe fires that can spread over larger areas (Westerling et al. 2006). The cumulative effects of these forces have reduced forest resilience by placing too much of the forest landscape at the same, vulnerable point in the fire cycle. The risks of catastrophic wildfire and disease outbreaks are high and climbing.

Case studies in the resilience literature point to three key features that help confer general resilience. Diversity is the first of these features; in the example of the Sierran forest landscape mosaic, diversity helps limit the size of disturbances and their impacts. Modularity, the second of these features, characterizes a system made up of coherent subsystems connected loosely enough that, when one fails, others can persist and rebuild the missing piece. A more resilient Sierran forest landscape would be managed to contain fires within discrete parts of the landscape, at small enough scales that surrounding forest stands can reseed burned areas. The third feature is the tightness of feedbacks, a measure of "how quickly and strongly the consequences of a change in one part of the system are felt and responded to in other parts" (Walker and Salt 2006). Whereas diversity and modularity largely maintain the independence of different system parts, tight feedbacks ensure sufficient connectedness and flow of information to enable adaptive responses to change, shocks, and surprises. In our Sierran example, resilience is enhanced by institutional or other mechanisms that link local events to larger-scale policy decisions, such as the feedbacks between state legislators and their constituencies provided by short legislative terms and the popular vote. As with spatial linkages, however, resilience is also constrained if feedbacks become so tight that they limit the flexibility of the system to respond to change. Resilient systems are characterized by feedbacks that are tight but not too tight (Walker and Salt 2006). Another perspective on this balance is that resilience calls for an appropriate balance between interconnectedness and independence across the landscape and across scales.

Resilience and Diversity

With native biodiversity at the core of today's park and wilderness values, the relationship between diversity and resilience deserves special attention. In the context of resilience, *diversity* refers broadly to landscape, institutional, and social features as well as to the native biodiversity. But resilience and native biodiversity have an especially interesting relationship because each is, at least in part, a means of enhancing the other. Resilience is a vehicle for sustaining native biodiversity in the long term, in part through its focus on sustaining processes such as the capacity for evolutionary adaptation and species range shifts. In turn, maintaining species and landscape biodiversity—keeping all the parts, as in Leopold's (1949) metaphor—can help foster resilience. In this sense, native biodiversity has both intrinsic value and a central role in the character of ecosystems and their options for the future. In working rangelands, maintenance of both rare and common plant species provides functional reservoirs. In years when some species that support a given function do poorly, other species that support that same function can step up to fill those roles (Walker et al. 1999). On serpentine outcrops in California, tiny endangered populations of drought- and heat-tolerant plants that today persist in hot, south-facing ravines provide insurance that something will spread to take the place of today's landscape dominants and the plant cover they provide if hot, dry conditions exclude them in the future (Stuart Weiss, personal communication, 2007). Maintaining diversity—even diversity that appears "redundant" (Gitay et al. 1996) in the short term ("we already have two canine predators; why do we need a third?")—is a strategy to hedge bets about the future and the limitations of our knowledge, to manage risk of undesirable functional changes when inevitable compositional and structural changes occur. The resilience of ecological and evolutionary processes and of native biodiversity is thus interwoven, with the important perspective shift that native biodiversity is dynamic and can acceptably undergo dramatic changes in distributions and relative abundances.

Ecological Resilience and Other Guiding Principles

How does the goal of ecological resilience differ from other management approaches, such as a focus on historical fidelity or autonomous nature? A goal of resilience emphasizes above all the capacity of an ecosystem to persist in the long term, within broadly defined, acceptable bounds of structure and function. On the surface, this articulation could seem very much like the goal of historical fidelity—to retain something into the future. The differences are matters of emphasis and scale, which turn out to be very important.

First is the "what" whose persistence is at issue. As described in Chapter 8, historical fidelity emphasizes *states*: composition, structure, iconic species, and their roles in creating a human experience of connectedness to the past in a protected area. In contrast, resilience emphasizes *functioning*: ecological processes and ecosystem capacity for adaptive response, often at scales larger than the protected area. A particular landscape mosaic of forests and meadows with its current species mix, such as that of Tuolumne Meadows in Yosemite National Park, might be an objective of historical fidelity. The ability of forest and meadow species to migrate and adapt to changing and variable climate and fire, reorganize into new assemblages as conditions shift, and resist invasion by transcontinental exotic species might be objectives more typical of resilience. The term *resilience* sometimes is also applied to specific communities, such as the sawgrass (*Cladium jamaicense*) marshes of Everglades National Park (Walker and Salt 2006) or the endemic-rich serpentine grassland patches and vernal pools of California, especially when species in them lack prospects for migrating elsewhere. However, it can be difficult or impossible to maintain the resilience of a system defined in narrow or small-scale terms if external stressors are changing directionally and cannot be addressed directly.

This issue of scale also helps define how resilience differs from historical fidelity. As discussed in Chapter 8, historic ranges of variability are often used to bound management objectives. Resilience focuses on a longer time perspective than the historic emphasis in the United States on ecosystems as they were when first encountered by Europeans. For instance, native biodiversity in a resilience framework is referenced to time scales of tens of thousands of years and the accompanying range in climate and other variables, over which species distributions moved individually and at scales much larger than single protected areas. From a resilience perspective, native biodiversity is a regional-to-continental, dynamic concept; where native caribou (*Rangifer tarandus*) and tundra reside today, native moose (*Alces alces*) and spruce (*Picea*) forests may reside in a century. This shift from local to regional and from a static to dynamic, process-centered perspective need not leave behind individual species—just shift the approach taken to secure their long-term persistence.

With respect to a hands-off management approach based on autonomous nature (Chapter 6), from a resilience perspective sustaining ecological dynamism and adaptability through active intervention will ultimately protect the autonomy of ecological systems more than will passively allowing their capacity to erode. In several parts of the world, nutrient-poor soils derived from pockets of mantle rock called serpentinite support unique, native-dominated, and endemic-rich ecosystems. In California, serpentine grasslands are threatened by exotic species invasions that appear to be aided by fertilization from automobile and industrial nitrogen emissions (Weiss 1999). Hands-off management, particularly near urban areas, would condemn these systems to low-diversity exotic dominance and losses of native species. Active management, involving cattle grazing, herbicide use, and prescribed fire, can reduce invasion, prevent extinctions and extirpations, and possibly reduce rates of nitrogen accumulation (personal observation and communications; Weiss 1999).

Perhaps even more than other guiding principles, resilience requires attention to human institutions and communities linked to protected areas. This is so for two reasons. First, protected areas are surrounded by working and inhabited landscapes with which managers must engage to protect large-scale processes that promote resilience, such as migration, gene flow, and species range shifts. Second, the long-term prospects of a protected area can be stronger if local communities support and participate in its conservation, or at least accept it, rather than oppose it, as is too often the case in North America. Promoting cultural connections to protected landscapes, as the U.S. National Park Service already pursues through avenues such as interpretation, is similarly important to their long-term persistence. The Resilience Alliance practitioner's workbook provides a useful guide to describing relevant ecological and social dynamics and can help pinpoint priority areas for additional information gathering (Resilience Alliance 2007).

What Managers Can Do about Resilience

To date, the concept of resilience has been used mainly to describe and elucidate the dynamics of social, ecological, and socioecological systems. Little specific guidance exists for managers seeking to increase and maintain resilience, although the Resilience Alliance (2007) recently released a workbook for practitioners interested in exploring the concept. Some scholars have taken the important step of describing what features of existing management systems are compatible with maintaining resilience in protected areas (e.g., Hughes et al. 2007). The further step of suggesting how to modify existing management systems to pursue a goal of resilience is a tricky one. Every protected area, region, and ecosystem is unique, requiring a tailored management approach rather than a one-size-fits-all prescription (Ostrom et al. 2007).

Rather than attempt to give specific instructions, we focus on four themes that we view as central to the goal of managing for resilience:

- Moving the basis of management from resistance toward resilience
- Defining what is to be sustained and what it should be resilient to
- Explicit integration of the regional and social context of protected areas
- Planning for the possibility that change will overwhelm a system's resilience

From Resistance to Resilience

Rapid and directional environmental changes make preservation of past and current landscapes, vistas, and ecological communities increasingly difficult. How can you balance retaining the system you hope to make resilient with letting it change for the sake of its resilience? One way to address this question is to ask whether the changes are reversible and whether resisting them increases the risk of irreversible effects. Another is to ask whether management can combine strategies, resist change in one setting, and promote it in another to keep options open.

For example, emblematic species will probably experience range shifts under climate change—in some cases, entirely out of the protected areas they symbolize. Joshua Tree National Monument might not naturally support Joshua trees (*Yucca brevifolia*), and Sequoia–Kings Canyon might no longer contain regenerating giant sequoias. What management approaches are consistent with maintaining or building resilience in such cases? In the very long term, protecting one type of resilience—the capacity of either park to return periodically to its twentieth-century state—entails maintaining the species in question somewhere within the region and maintaining connectivity between old and new distributions. Maintaining another component of resilience—the sustained functioning of the protected area through maintenance of living native vegetation in situ—probably entails allowing other native species to expand into the protected area to replace the declining flagship species.

To allow for adaptive change, managers probably would not want to remove seedlings of other tree species encroaching into giant sequoia groves or Joshua tree stands in an effort to reduce competition. They might begin identifying and working to protect areas where climate might suit giant sequoias over the next few centuries. They might also decide to pursue a resistance strategy: to maintain, through labor-intensive watering, selected stands of Joshua trees or sequoias in the protected area. Without the resilience piece, the persistence of each species might come to depend entirely on costly annual watering for decades or centuries, a tenuous strategy to prevent irreversible species loss. Without the resistance piece, public perception of management might sour as they perceive neglect of the park's iconic species. The resistance piece also hedges against the possibility that Joshua trees or giant sequoias cannot be effectively protected or might fail to establish elsewhere. The most resilient approach is diverse. It considers the role of the public and includes options, above all, to spread risk and prevent irreversible species losses. Chapter 13 describes tools such as scenario-based planning that managers can use to identify robust strategies to promote resilience in cases such as these.

Resilience of What to What?

Although park managers want resilience to the unexpected (general resilience), from a practical perspective resilience can be defined and planned mainly with respect to the likely and possible specific stresses that protected areas face (Resilience Alliance 2007). Depending on location, these stresses might include climate change, extreme events such as hurricanes or floods, fire ignitions, intensive human use, and species invasions. Protected areas also need to be resilient to specific stresses outside the ecological realm, such as changing values, technology, laws, and economic pressures. Finally, disturbances do not need to be novel to challenge the resilience of a protected area. Ecosystems also need to be resilient to historically important disturbances, such as fire or flooding, particularly if these regimes have changed in one or more dimensions such as frequency, duration, severity, or predictability (Resilience Alliance 2007). Novel combinations of old and new disturbances can also interact in new ways that make a previously tolerable disturbance newly threatening. History, current events, model projections, and traditional knowledge are among the diverse sources that managers can draw on to identify and prioritize current, likely, and possible stressors to which a system should be resilient.

Defining what about a system should be resilient could fall to the public, scientists, managers, or policymakers; deciding who decides is not trivial. Prevailing global protected area priorities suggest that native biological diversity should be a target of resilience and that its definition must include genes, species, populations, and biomes not currently within the boundaries of a protected area that might spread into it in response to environmental changes. Core functional attributes of biodiversity should probably also be targeted, such as native plant cover sufficient to stabilize soils and provide a variety of habitats at landscape and larger scales. Structural attributes of the system and its diversity are a third likely target, such as complete or complex food and species interaction webs. Finally, explicit human values, such as cultural connections to the land or a particular quality and type of experience on the landscape, are important targets of resilience. A list of what is to be resilient is central to managers' ability to navigate trade-offs in decision making and to measure the success of their strategies.

How can one assess the resilience of a protected area or ecosystem before something catastrophic happens to test it? As described in Chapter 13, a sustainable monitoring program for a well-defined, small set of indicators is the critical ingredient in detecting stresses and changes in them, ecosystem responses to them, and signs that directional stressors are producing directional trends in the ecosystem that might have important cumulative or threshold effects. One type of proxy or benchmark for resilience is a small set of key processes that can change only gradually but that strongly shape the functioning and character of an ecosystem. For instance, sediment phosphorus levels in lakes and wetlands change gradually in response to surrounding land use and govern dominant vegetation, food web complexity, and the shift between oligotrophic and eutrophic states (Carpenter et al. 2001; Walker and Salt 2006). In terrestrial ecosystems, key variables might include soil organic matter content (Chapin et al. 2004), biologically available soil nitrogen, or the presence of particular functional types of plants and animals in the regional species pool (Chapin et al. 2006). By monitoring well-chosen "slow variables," managers can detect changes, or accelerations in rates of change, that signal likely declines in resilience by pushing the ecosystem closer to a qualitative change in character such as a shift from grassland to scrubland. Identifying which of these slow variables to monitor depends on what processes, ecosystem services, or characteristics management seeks to sustain. The Resilience Alliance practitioner's workbook (2007) provides a useful guide to begin identifying appropriate slow variables to target for monitoring in the system of interest.

Protected Areas in Regional Context

Resilience as a concept explicitly embeds protected areas in their broader geographic regions and social contexts. In the ecological realm, management should address the porous nature of boundaries between protected areas and surrounding lands. Wildlife, exotic species, fire, disease outbreaks, contaminants, and water move across these boundaries; managing them effectively requires coordination with landowners and managers of other jurisdictions. In some areas, bridging institutions already exist to facilitate this region-scale coordination. For example, the purpose of the Greater Yellowstone Coordinating Committee (2008) is to link the National Park Service, U.S. Forest Service, and U.S. Fish and Wildlife Service to coordinate federal land management in the Montana–Wyoming Yellowstone region. With the added directional push of climate change, management to permit movement through larger landscapes will become even more important to maintain species and the functional roles they play (Heller and Zavaleta 2009).

In the social realm, protected area management with a goal of resilience should consider the roles of regional economic, cultural, and community trajectories in the long-term sustainability of the park and its biodiversity. Local people and their commitment to natural places should be a cornerstone of long-term protection, but in North America local communities and protected areas often are at odds. In some regions, such as parts of Alaska, local people have the historical experience with a place and the vested interest to want to steward it well. In other places, a focus on economic gain through extraction and development has clashed with protected area stewardship. Managing for long-term, regional resilience includes maintaining and rebuilding people's long-term relationships with and commitments to wild nature and to the larger landscapes in which protected areas exist.

In the institutional realm, resilience can be an effective guiding principle at larger scales than individual protected areas. Systems of protected areas, such as the National Wildlife Refuge system or the U.S. system of national parks, and the agencies that manage them can be more or less able to adapt, absorb change, and persist as effective entities. At this larger scale, one once again needs to ask what should be resilient—for example, the capacity of the system to protect biological diversity in and around individual protected areas or its positive perception by the public in regions that contain protected areas. Cross-institutional bridges (the same balance of independence and interconnection) can help sustain the integrity of protected areas by integrating management along a continuum from urban to reserve areas and addressing issues such as compact urban design and zoning extractive uses outside protected areas. This embedded approach is especially important in landscapes where protected areas are small but processes that maintain resilience, such as fire and predation, require large landscapes.

Planning for the Possibility of Transformation

Can or should management always focus on maintaining the resilience of the current ecosystem? Rapid, directional change can overwhelm the capacity of even a highly resilient ecosystem to adapt. Coral reef ecosystems vary in resilience depending largely on how heavily they are fished and how much nutrient pollution they experience (Hughes et al. 2007). Their ability to recover from bleaching events—episodic coral die-offs driven by unusually warm temperatures—suffers as fertilization and fish declines increase the growth of algae that compete with corals to recolonize reef sites (Bellwood et al. 2004; Hughes et al. 2007). Climate change challenges the managers of coral ecosystems to mitigate these stressors and to encourage shifts toward more warm-adapted coral species, among other possible actions. But in at least some ocean regions, ocean acidification threatens to overwhelm even healthy reef ecosystems by impeding the coral's ability to pull dissolved calcium carbonate out of the water column and lay it down in the skeletons that build the reef (Hoegh-Guldberg et al. 2007). If no amount of resilience can overcome this fundamental shift in ocean chemistry, then what? Managers face three options: to resist change to the death, watch the system passively degrade, or attempt to actively transform it into something entirely new that delivers at least some of the desired functions of the former ecosystem (Chapin et al. 2006).

What might this new system look like, and how would transformation become possible at institutional and ecological levels? Although managers may rightly hope it never comes to that, planning for the possibility of transformation makes that option available in case some shock or change overwhelms the current system's resilience. Planning for when to shift strategies from resilience to transformation—at what thresholds, with what level of information—also helps overcome the tendency to stick with old management goals and strategies long after they have ceased to fit reality on the ground. For instance, a shift in strategy from managing for in situ conservation of Joshua trees in southeastern California to also actively assisting their movement to new locations should begin, ideally, before mass die-offs in situ have occurred. Practical guidance for managers about when to shift strategies from resilience within the current system to transformation toward a new state is actually beginning to emerge in light of challenges posed by climate change (see Hoegh-Guldberg et al. 2008).

Sustaining Protected Area Values through Resilience in Practice

The specter of irreversibility hangs over protected area stewardship. Managers can only fail once to prevent an extinction, which is truly forever, or a dramatic and long-term state shift such as dominance by invasive plants. In a time of tremendous directional change in the biophysical, cultural, and socioeconomic realms, sustaining protected area values for the long haul should be an overarching principle, regardless of whether interventions are needed to achieve them. We argue that core values, including native biodiversity, ecosystem services, and cultural connections to the land, cannot be sustained without resilience: the capacity of a system to weather disturbances and surprises and to self-organize in response to perturbations without a fundamental loss of character. Ultimately, successful management for resilience should decrease the need for intervention and intensive management. As resilience grows, so does an ecological system's capacity to adapt on its own to shocks and change.

Opportunities are growing to deploy resilience as a guiding principle for protected area management. This is timely, given that trends in most of the world mean the need is growing to deploy resilience and long-term sustainability as core goals for protected areas. Available tools include the pursuit of resilience science in protected areas such as South Africa's Kruger National Park (Venter et al. 2008), the development by the Resilience Alliance of ecological case studies for U.S. protected areas such as the Florida Everglades (Walker and Salt 2006) and a guide for managers (Resilience Alliance 2007), and the growing set of planning resources for managers to tackle climate change and uncertainty (see Chapter 11). Other tools include enabling legislation, such as the National Wildlife Refuge System Improvement Act of 1997, which directs managers to use both voluntary and legally mandated means to expand stewardship beyond the boundaries of refuges and into larger surrounding landscapes (Meretsky et al. 2006).

Managing for resilience ultimately means managing for the long-term adaptability and functioning of a regional system, even if that means allowing major reshuffling of the ecosystem's parts to take place (without losing parts altogether). Box 9.1 summarizes the strategies described in this chapter and organizes them in terms of four guiding principles. Sustaining an ecosystem over time in a rapidly changing world entails reducing its vulnerability to shocks; building its ability to adapt to change; increasing the resilience of the ecosystem in its current, broadly defined state; and sometimes taking steps to facilitate transformation to a new resilient state in order to avert catastrophic collapse in the face of stresses. The practice—as opposed to the theory—of resilience is young, but the time is ripe to learn how to cultivate adaptable, self-sustaining protected area ecosystems. The push for maturity will probably require managers to take the reins, because ultimately only they can really explore the practical matter of how to build resilience—and to measure how much of it exists—in our rapidly evolving protected areas.

BOX 9.1. STRATEGIES TO INCREASE THE RESILIENCE OF DESIRED SYSTEMS

- Reduce exposure and sensitivity to stresses (vulnerability; see Turner et al. 2003) by
    Sustaining resources (slow variables)
    Mitigating stresses that drive change
    Reducing sensitivity to stress by fostering ecosystem capacity to absorb stresses and shocks, such as through the maintenance of buffers and redundancy
- Increase adaptability by
    Fostering ecological, economic, and cultural diversity (the building blocks for change)
    Fostering learning and innovation at multiple scales
- Increase resilience of the current system state and its regional context by
    Strengthening and stabilizing feedbacks
    Fostering cultural and ecological memory and connections, such as reconnecting local and urban people with the land
    Building linkages across scales (connectivity, adaptive governance) by managing the matrix and protected areas and by nesting site management within a regional strategy
- Foster transformability by
    Thinking creatively about what desirable, functioning ecosystems a protected area can sustain under new or changing conditions
    Strategizing beyond the protected area about how to keep all the parts (biodiversity, ecosystem processes and services) during and after transformation
    Treating crisis as an opportunity for constructive change
    Deciding ahead of time what will trigger an intentional shift to a strategy of transformation

REFERENCES

Bellwood, D. R., T. P. Hughes, C. Folke, and M. Nystrom. 2004. Confronting the coral reef crisis. *Nature* 429:827–833.

Carpenter, S., B. Walker, J. M. Anderies, and N. Abel. 2001. From metaphor to measurement: Resilience of what to what? *Ecosystems* 4:765–781.

Chapin, F. S., A. L. Lovecraft, E. S. Zavaleta, J. Nelson, M. D. Robards, G. P. Kofinas, S. F. Trainor, G. D. Peterson, H. P. Huntington, and R. L. Naylor. 2006. Policy strategies to address sustainability of Alaskan boreal forests in response to a directionally changing climate. *Proceedings of the National Academy of Sciences of the United States of America* 103:16637–16643.

Chapin, F. S. III, G. Peterson, F. Berkes, T. V. Callaghan, P. Angelstam, M. Apps, C. Beier, et al. 2004. Resilience and vulnerability of northern regions to social and environmental change. *Ambio* 33:344–349.

Gitay, H., J. B. Wilson, and W. G. Lee. 1996. Species redundancy: A redundant concept? *Journal of Ecology* 84:121–124.

Greater Yellowstone Coordinating Committee. 2008. Greater Yellowstone Coordinating Committee Web site. Retrieved October 12, 2009 from fedgycc.org/gycc_roles.htm.

Heller, N. E., and E. S. Zavaleta. 2009. Biodiversity management in the face of climate change: A review of 22 years of recommendations. *Biological Conservation* 142:14–32.

Hoegh-Guldberg, O., L. Hughes, S. McIntyre, D. B. Lindenmayer, C. Parmesan, H. P. Possingham, and C. D. Thomas. 2008. Assisted colonization and rapid climate change. *Science* 321:345–346.

Hoegh-Guldberg, O., P. J. Mumby, A. J. Hooten, R. S. Steneck, P. Greenfield, E. Gomez, C. D. Harvell, et al. 2007. Coral reefs under rapid climate change and ocean acidification. *Science* 318:1737–1742.

Holling, C. S. 1973. Resilience and stability of ecological systems. *Annual Review of Ecology and Systematics* 4:1–23.

Hughes, T. P., D. R. Bellwood, C. S. Folke, L. J. McCook, and J. M. Pandolfi. 2007. No-take areas, herbivory and coral reef resilience. *Trends in Ecology and Evolution* 22:1–3.

Leopold, A. 1949. *A Sand County almanac*. Oxford University Press, Oxford.

Mack, R. N. 1981. Invasion of *Bromus tectorum* L. into western North America: An ecological chronicle. *Agro-Ecosystems* 7:145–165.

Meretsky, V., R. L. Fischman, J. R. Karr, D. M. Ashe, J. M. Scott, R. F. Noss, and R. L. Schroeder. 2006. New directions in conservation for the National Wildlife Refuge system. *BioScience* 56:135–143.

Millennium Ecosystem Assessment. 2005. *Ecosystems and human well being: Status and trends*. Island Press, Washington, DC.

Ostrom, E., M. A. Janssen, and J. M. Anderies. 2007. Going beyond panaceas. *Proceedings of the National Academy of Sciences* 104:15176–15178.

Pergams, O. R. W., and P. A. Zaradic. 2008. Evidence for a fundamental and pervasive shift away from nature-based recreation. *Proceedings of the National Academy of Sciences of the United States of America* 105:2295–2300.

Resilience Alliance. 2007. *Assessing and managing resilience in social-ecological systems: A practitioners workbook*. Volume 1, version 1.0. Retrieved October 12, 2009 from www.resalliance.org/3871.php.

Resilience Alliance. 2008. *Resilience*. Retrieved October 12, 2009 from www .resalliance.org/576.php.

Sierra Nevada Ecosystem Project. 1996. *Status of the Sierra Nevada: Summary of the Sierra Nevada ecosystem project report*. University of California Centers for Water and Wildland Resources, Davis.

Turner, B. L., R. E. Kasperson, P. A. Matson, J. J. McCarthy, R. W. Corell, L. Christensen, N. Eckley, et al. 2003. A framework for vulnerability analysis in sustainability science. *Proceedings of the National Academy of Sciences of the United States of America* 100:8074–8079.

Venter, F. J., R. J. Naiman, and H. C. Biggs. 2008. The evolution of conservation management philosophy: Science, environmental change and social adjustments in Kruger National Park. *Ecosystems* 11:173–192.

Vitousek, P. M., H. A. Mooney, J. Lubchenco, and J. M. Melillo. 1997. Human domination of Earth's ecosystems. *Science* 277:494–499.

Walker, B., A. Kinzig, and J. Langridge. 1999. Plant attribute diversity, resilience, and ecosystem function: The nature and significance of dominant and minor species. *Ecosystems* 2:95–113.

Walker, B., and D. Salt. 2006. *Resilience thinking: Sustaining ecosystems and people in a changing world*. Island Press, Washington, DC.

Weiss, S. B. 1999. Cars, cows, and checkerspot butterflies: Nitrogen deposition and management of nutrient-poor grasslands for a threatened species. *Conservation Biology* 13:1476–1486.

Westerling, A. L., H. G. Hidalgo, D, R. Cayan, and T. W. Swetnam. 2006. Warming and earlier spring increase western US forest wildfire activity. *Science* 313:940–943.

PART III

Management Strategies for Implementing New Approaches

The first part of this book built an argument for why the concept of naturalness is insufficient as a guide for making decisions about where, when, and how to intervene in park and wilderness ecosystems. More specific and diverse goals are needed, goals that can be translated into objectives and implemented in management practice. The importance of diverse goals reflects the high level of uncertainty about the future; as global change accelerates, diversification is a means of hedging risk. Diversity also is an effective response to park and wilderness purposes being varied and, to some extent, in conflict with each other. The second part of this book described four different management goals or approaches that park and wilderness managers might use, each with a different resultant emphasis and outcome.

This third part of the book delves more deeply into the practical strategies and specific actions protected area managers might consider in the context of new goals and approaches, dynamic ecosystems, and global change. The public land management agencies in the United States have a proud tradition of stewardship, with time-honored institutions and programs and with highly devoted and knowledgeable staff. Nevertheless, as mentioned in earlier chapters, challenges to these institutions, programs, and knowledge base are mounting at an ever-increasing pace. Past practice may be ineffective in the future. The goal of the chapters in Part III is to suggest specific ways in which current approaches might be adapted to meet the challenges of twenty-first-century conservation.

As threats diversify and expand and more ecosystems and ecological elements are endangered, it has become impossible to protect all values from all possible threats in all places. Choices must be made. Chapter 10 emphasizes prioritization and the importance of developing realistic, attainable objectives, using the example of invasive species. In the long fight to control nonnative invasive species, it has become clear that all invasive species cannot possibly be eliminated, that different species constitute different kinds of threats, and that some responses are more effective than others. With limited resources, decisions must be made about where, when, and how to fight invasive species. Management of invasives therefore provides practical lessons about the need to develop clear, tangible conservation goals and objectives before threats can be assessed and addressed. The work on invasives also demonstrates how to prioritize threats and strategies and how to use triage to determine where to focus limited resources.

Chapter 11 explores potential management responses to climate change as a means of illustrating the importance of a toolbox approach to management, implementing diverse approaches that vary in temporal and spatial scale. A consistent theme is the need to acknowledge uncertainty, to learn, and to be prepared to adapt and change course as new information accrues or surprises occur. Depending on the situation, both reactive and proactive approaches can be appropriate. Existing management programs can be adjusted to account for climate change, or nontraditional approaches can be attempted, ideally in a cautionary, experimental manner that allows for learning.

The importance of hedging bets is one of the themes of Chapter 11—indeed, of this entire book. One way to hedge bets and spread risk is through planned implementation of diverse management approaches at large spatial scales. Chapter 12 explores planning and management at large spatial scales in more detail. At least three interrelated types of large-scale planning and coordination are needed to conserve biodiversity and meet protected area goals: coordination between separate protected areas; planning of the spatial configuration of protected areas into networks, nodes, and corridors; and integrated management across the landscape, managing protected areas and the matrix in which they are situated. Barriers to large-scale planning and management are profound. Achieving conservation at large scales requires the development of policies and institutions that do not currently exist.

Building on earlier chapters that focus on dealing with uncertainty (Chapter 4 on global change, Chapter 9 on resilience, and Chapter 11 on managing climate change), Chapter 13 explores ways in which protected area planning might evolve to better respond to the conservation challenges of the twenty-first century. To effectively deal with rapid change, uncertainty, and surprise, planning must become more adaptive—more dynamic, flexible, and responsive. Tools such as sensitivity analysis and scenario planning can help planners and managers make decisions despite uncertainty. Processes that engage the public and stakeholders in meaningful dialogue about the future of protected areas are critical.

The final chapter in this part, Chapter 14, promotes the notion of wild design. It argues that when we intervene, we should acknowledge that we interject human intention into wild ecosystems, designing them. Once this is acknowledged, it is important to think carefully about an ethical approach to design that is appropriate for parks and wilderness ecosystems. The chapter articulates a set of principles intended to provide a starting point for developing wild design into a full-fledged practice for protected area interventions.

Chapter 10

Objectives, Priorities, and Triage: Lessons Learned from Invasive Species Management

JOHN M. RANDALL

We are living in a period of the world's history when the mingling of thousands of kinds of organisms from different parts of the world is setting up terrific dislocations in nature.

—*Charles Elton*

Nonnative species can establish and spread in protected areas, and some can be drivers of damaging and undesirable change to native species, populations, and communities, as well as to biological and ecosystem processes. I will use the term *invasive species* to refer to the subset of nonnative species that grow and spread beyond cultivation or other human uses and cause or have the potential to cause harm to native biodiversity, the economy, or human health, following the definition adopted by the U.S. National Invasive Species Council (U.S. Department of Agriculture 1999). Invasive species were recognized as a threat to nature conservation and native species at least as early as the mid-1800s. For example, within a year of the 1864 establishment of Yosemite as a park, concerns about weeds choking out native vegetation were raised by the commission appointed to manage the valley (Olmsted 1865). Established invaders continue to spread, and many new invasions continue to occur in terrestrial, freshwater, and marine environments around the world. A study by the International Union for Conservation of Nature found that invasive species have been reported in protected areas in 106 countries (DePoorter et al. 2007). Even remote areas such as designated wilderness and uninhabited sub-Antarctic islands are plagued by invasive species. Efforts to prevent and fight these invasions have yielded both notable successes and striking failures, which provide lessons that apply to an array of anthropogenic threats that challenge protected area managers.

Earlier chapters in this book argued that the concept of naturalness, as a goal of protected area stewardship, is vague, imprecise, and unrealistic. In contrast, goals such as the preservation of ecological integrity or of native biological diversity provide a better framework for making difficult management decisions, particularly where such decisions involve ecosystem interventions and other aggressive actions. The challenges of managing invasive species make it clear that the identification of tangible focal targets and conservation objectives is critical to management success. Where this has not been done, resources have been wasted and, in some cases, eradication and control efforts have done more harm than good (Rinella et al. 2009).

This chapter begins with a brief overview of the effects of invasive species on biological diversity in protected areas. Two interrelated lessons learned from efforts to address invasive species are then described. First, it is necessary to develop clear, tangible conservation goals and objectives before threats to them can be assessed and addressed. Second, it is necessary to set priorities among these threats and among strategies for addressing them. Drawing from The Nature Conservancy's experience in preventing and managing invasive species, this chapter illustrates how lessons from invasive species management apply to many different anthropogenic threats to protected areas.

Impacts of Invasive Species on Biological Diversity

Invasive plants, animals, fungi, and microbes threaten native biological diversity in ecosystems around the world. Collectively, they are often ranked second only to habitat destruction as a threat to biological diversity worldwide (Wilson 1992). In 1994, 61 percent of 246 national park superintendents responding to a survey indicated that nonnative invasions of plants were moderate or major problems in their parks (Layden and Manfredo 1994). Invasive species were the most frequently cited threat in a 2009 summary of 974 Nature Conservancy projects around the world, listed by 587 (60 percent) of the projects (unpublished data, The

Nature Conservancy). According to Wilcove et al. (1998), invasive species contributed to the imperilment of nearly half of the plants, birds, and fish considered rare, threatened, endangered, or extinct in the United States.

Negative effects of invasive species on biological diversity and ecological integrity may include alteration of ecosystem processes, change in community composition and structure, suppression or elimination of native species populations, and genetic effects. An invader's effects may change over time in intensity, character, or direction (Strayer et al. 2006). For example, the red fire ant (*Solenopsis invicta*) reduced populations of other insects when it first established and became abundant in a Texas invasion, but 12 years later, populations of native insects had returned to pre-invasion levels (Morrison 2002).

The most damaging invasive species alter ecosystem processes. For example, cheatgrass (*Bromus tectorum*) has invaded millions of hectares in North America's intermountain West, where it is responsible for increasing fire frequencies from once every 60 to 110 years to once every 3 to 5 years (Whisenant 1990). Bradley et al. (2006) concluded that the expansion of cheatgrass in the Great Basin alone has released an estimated 8 (± 3) teragrams (10^{12} grams) of carbon to the atmosphere and that it will probably release another 50 (± 20) teragrams of carbon in the coming decades, changing portions of the western United States from a carbon sink to a carbon source. Invasive animals can drive ecosystem-level changes too. Predation by arctic foxes (*Alopex lagopus*) introduced to Aleutian islands extirpated or greatly suppressed large seabird colonies, which led to sharp decreases in deposition of guano, rich in nitrogen and other plant nutrients, and ultimately to changes in the vegetation, from lush grasslands to far less productive dwarf shrub and forb-dominated communities (Croll et al. 2005).

Invasive diseases and pests that attack native species can have devastating impacts. In the first half of the twentieth century, the introduced chestnut blight fungus (*Cryphonectria parasitica*) killed more than 3 billion American chestnut trees (*Castanea dentata*) in eastern North America, virtually eliminating a species that had dominated large areas of deciduous forest. Competitive displacement of native species by invasives is often suspected but surprisingly has rarely been demonstrated. One example was documented by Bøhn et al. (2008), who found that the invasive fish vendace (*Coregonus albula*) competed for food with the densely rakered form of the native whitefish (*Coregonus lavaretus*), which was originally dominant in pelagic habitat in the subarctic Pasvik watershed, shared by Finland, Norway, and Russia. This led to a 90 percent decline in the native's population density over the 14-year study.

Invasive species can also eliminate or sharply reduce the frequency of native genotypes and specific alleles. The most striking examples of this involve hybridization between an invasive and a related native species, potentially leading to the complete elimination of "pure" native individuals. Where more than one invasive species are present, they may have strong, sometimes mutualistic interactions that can amplify their effects on biodiversity. Simberloff and Von Holle (1999) called this positive feedback process invasional meltdown. For example, on the island of Hawaii, nitrogen fixation by the invasive fire tree (*Myrica faya*) and its actinorrhizal symbiont, *Frankia* sp., promotes populations of nonnative earthworms, which in turn increase rates of nitrogen burial, further altering soil nutrient cycling in these systems (Aplet 1990).

Some introduced species may have net positive effects on native biodiversity in some locations and therefore are not considered invasive. In some cases introduced species can provide a limiting resource, increase habitat complexity, functionally replace a native species, or control another, more damaging invasive species (Rodriquez 2006). Biological control agents of invasive species are almost always nonnative themselves but are released with the expectation that they will do more good than damage (Figure 10.1). Follow-up monitoring has revealed this to be the case in many instances (e.g., Lesica and Hanna 2004).

FIGURE 10.1. This nonnative weevil (*Larinus minutus*) was deliberately introduced to reduce the spread of spotted knapweed (*Centaurea stoebe*). (Photo by Paul Alaback)

Lessons Learned

The successes and failures resulting from decades of efforts to prevent and control invasive species in protected areas have yielded important lessons. Two interrelated lessons are most important for improving the ways in which invasive species threats to biodiversity are addressed: the need to identify and clearly articulate conservation goals and objectives for a protected area and the need to prioritize threats and the strategies for addressing them.

Identifying and Clearly Articulating Conservation Goals and Objectives

Decades of experience have shown that it is necessary to develop and clearly articulate protected area goals and objectives before invasive species threats can be adequately assessed and strategies to address them developed and implemented. If it is not clear what is being protected, it will be difficult to determine what is threatening it. Articulating protected area goals and objectives entails identifying the objects and processes of most concern, the so-called target species, populations, communities, and processes that the protected area's managers should be managing for. This, in turn, allows planners and managers to identify the most severe threats to those targets so they can avoid wasting time and resources on conspicuous but minor threats. It also allows them to assess the relative importance and vulnerability of different places within the protected area.

For the past 100 years or so, maintaining naturalness was the stated goal of protected area management on public lands in the United States. As discussed in earlier chapters, naturalness is an imprecise goal, which includes an array of sometimes incompatible meanings and, as a result, provides vague and sometimes contradictory direction on how to assess and address invasive species and other threats. The more specific conservation objectives proposed in Chapters 7 through 9 generally offer more useful and practical guidance. In contrast to an emphasis on protecting wildness, as described in Chapter 6, hands-on actions to control and mitigate the impact of invasive species introduced and spread by humans may be an appropriate case of two wrongs making a right.

Clarity and practicality are both possible if the goal and the specific management objectives are based on protecting native biodiversity or ecological integrity, and it is here that I will focus most of the rest of my discussion. According to these goals, invasive species are evaluated on the basis of their effects on populations and communities of native species and on important biotic and abiotic factors and processes that sustain them. Invaders that substantially degrade biological diversity and ecological integrity are high-priority targets for prevention and control, whereas those with less effect are lower priorities.

Over the past 15 years, The Nature Conservancy and partner organizations have developed a useful and robust planning process, Conservation Action Planning (CAP). This process gives protected area planners and managers a powerful tool for determining and articulating conservation goals, objectives, and priorities and for developing strategies to achieve those goals, as well as a means to measure progress toward them and then to adapt and learn over time (The Nature Conservancy 2007). CAP has four basic steps. First, identify goals and priorities, including the targeted species, populations, communities, and processes; second, identify threats to the target biodiversity and develop strategies to address those threats and protect target biodiversity; third, implement the strategies; and fourth, measure the effects of the actions taken and use the results to adapt and improve (The Nature Conservancy 2007). The rigorous and systematic CAP process has similarities with the way Parks Canada has operationalized ecological integrity, as described in Chapter 7.

The first step involves selecting a minimum set of conservation targets, comprising species, communities, and ecological systems that are chosen to represent and encompass the biodiversity found in the protected area (usually eight or fewer). Application of CAP suggests that conservation of the focal targets can ensure conservation of the full suite of native biodiversity within a functional landscape. For The Nature Conservancy, selection of focal conservation targets is often informed by ecoregional conservation priorities, which are the species and communities deemed most characteristic and important to preserve the full biodiversity of an ecoregion. This step also includes an assessment of the viability of each focal target, including a determination of key ecological attributes necessary for their long-term viability. For example, fire frequency might be identified as a key ecological attribute of a grassland community target.

The second step involves identifying threats to each of the focal targets and their associated ecological attributes, as well as a structured approach for ranking these threats as very high, high, medium, or low. For example, invasive predators (rats and foxes) are a very high threat to seabirds on the Aleutian Islands and other islands in the Bering Sea (conpro.tnc.org/438/project_info). In the Laramie Foothills/Phantom Canyon Preserve area in northern Colorado, the invasive plants leafy spurge (*Euphorbia esula*), knapweeds (*Centaurea* spp.), and toadflax (*Linaria dalmatica*) threaten at least four focal target communities (conpro.tnc.org/672/project_info).

This second step also involves developing strategies to prevent and abate threats and to protect focal conservation targets. This entails setting objectives in specific and measurable terms to describe what success looks like. Objectives can be based on the abatement of threats, the restoration of degraded key ecological attributes, or other outcomes of specific conservation actions. Strategic actions to achieve these objectives can then be selected and planned. For example, objectives for the Aleutian Islands include eradicating introduced predators and grazers from five outer Aleutian Islands and ensuring that on-the-ground rat prevention response occurs within 12 hours of all actual and potential boat groundings. At the Laramie Foothills site, one of the objectives is to control the extent of high-priority invasive species to less than 3 percent of the landscape and no patch greater than 0.5 hectares.

Where invasive species have been identified as major threats, strategies and actions should be selected based on their potential to prevent and control the invader or its harmful impacts on the target without causing greater harm to any of the targets. Such actions may cause some damage to vegetation or kill individuals of native species, but overall they should yield significantly greater benefits than harm to the focal conservation targets and to native biodiversity.

Prioritizing Threats and the Strategies to Address Them

Once invasive species have been identified as significant threats, it is necessary to prioritize different invasive species (or species groups), sites within the protected area, and strategies for preventing and controlling the invaders and restoring or rehabilitating sites they have damaged. The CAP process includes a structured method for ranking threats and distinguishing introduced species that do not pose threats to native biodiversity and focal targets from those that do (Zavaleta et al. 2001). Some protected areas use structured processes, such as the Alien Plant Ranking System (APRS), to prioritize the effects of different invaders or even different populations of a given invasive species (APRS Implementation Team 2000). The APRS focuses exclusively on plants and does not pay attention to invasive species established nearby and poised to invade, species that should be given highest priority if they are also capable spreading to the protected area and causing damage.

PRIORITIZING INVASIVE SPECIES THREATS

The following factors should be considered in prioritizing invasive species threats: current and potential effects of the species on native biodiversity (focal conservation targets), current distribution and abundance on or near the site, trends in distribution and abundance (e.g., rate of spread and increase in abundance), difficulty of control, and biodiversity value of the sites the species occupies or may occupy. The first two factors equate with the formula proposed by Parker et al. (1999), which calculates the impact of a nonnative species as the product of its per capita impact, its mean abundance in the area it occupies, and its range. The fifth factor is of particular importance for prioritizing invasive species at the scale of a protected area, which may contain sites and habitats with greater and lesser biodiversity value.

In general, the highest priorities are given to species that have large harmful impacts (e.g., those that change ecosystem properties such as fire frequency or that significantly alter community structure and composition); those that have current ranges and abundances that are small relative to their potential and rapid rates of spread and increase in abundance; those that are not difficult to control; and those that occupy areas with high biodiversity value. Usually the greatest weight is given to the current and potential effects of the species on native biodiversity, but it may be difficult to determine how even well-known species will affect biodiversity on a given site. The current extent of the species and its rate of spread are important because it is usually most efficient to devote resources to preventing new invasions and immediately addressing incipient infestations (Moody and Mack 1988). On the other hand, Ricciardi and Cohen (2007) found that the rates of establishment and spread of introduced species are not correlated with their impact on native species and biodiversity.

The Cosumnes River Preserve in central California offers an example of how priorities can be set among invasive species. Invasive plants were identified as threats to the vernal pool grasslands, floodplain, and blue oak (*Quercus douglasii*) woodland focal target communities. The preserve's 2001 weed management plan prioritized more than fifteen invasive plant species using criteria much like the five factors listed earlier (Meyers-Rice and Tu 2001). High priority was assigned to preventing invasions by Chinese tallow (*Sapium sebiferum*) and scarlet wisteria (*Sesbania punicea*), species known to be invasive in nearby watersheds but not yet present on the preserve. High priority was also assigned to several other invasive tree species such as edible fig (*Ficus carica*) and tree of heaven (*Ailanthus altissima*), which were present in low numbers on the preserve but judged capable of spreading rapidly unless quickly eradicated or contained. On the other hand, yellow starthistle (*Centaurea soltsitialis*) and Himalayan blackberry (*Rubus armeniacus*) were assigned low priority because the former occupies portions of the preserve that have low biodiversity value, and the latter was judged to have moderate impacts on targets and to be unlikely to become more widespread or abundant than it already was.

PRIORITIZING SITES

Prioritizing sites within a protected area may also be necessary. For example, the importance of tackling new outlier populations, the so-called nascent foci, of invasive plants before attempting to control larger source populations has become widely accepted in many situations (Moody and Mack 1988). For the 150,000-hectare Centennial Valley in southwestern Montana, spatially explicit simulation of weed spread revealed that a strategy focused on early detection and rapid control of new outbreaks was the most effective alternative (Martin et al. 2007). However, such rules do not always apply. Wadsworth et al. (2000) found that this strategy was likely to be futile for plants that spread long distances rapidly, such as some riparian invaders whose propagules are carried long distances downstream. For such plants and other invaders that spread rapidly, controlling source populations first can be most effective, especially if they are in upstream or upwind locations advantageous for long-distance dispersal.

PRIORITIZING STRATEGIES

Once invasive species threats have been prioritized, strategies to address the most important threats must be developed and prioritized. Objectives of these strategies may be to prevent an invasion or the spread of the invasive species; eradicate, contain, or control the abundance of the invasive species; mitigate the harmful impact of the invasive species; or restore or rehabilitate affected native species, communities, and processes. The amount of money and other resources available will also be important in determining priorities because they will limit what can realistically be done (Shea et al. 2002). Strategies that promise the greatest long-term protection of biodiversity and reduction in threat per unit of effort expended are favored. Prevention strategies should be assessed for their likely outcome relative to the damage deemed likely to occur if, without intervention, new species invaded and spread.

Strategies to Prevent Invasions

Strategies to anticipate and prevent new invasions deserve far more attention than they are typically given. Policies and programs to prevent new introductions and the spread of invasions are often carried out at national, regional, or state scales, but when successful they provide great benefit to protected areas. Protected area managers have an important role to play in advocating such policies because they can present compelling evidence of the damage caused by invasive species and examples of the benefits that stronger policies would yield. Managers can also implement strategies to prevent invasions of the lands and waters directly in their charge, by identifying potential invaders and pathways by which they might invade and then taking actions to close them. Prevention strategies of this sort are likely to be effective only when the pathways are readily subject to control, as when they rely largely or completely on human activities that can be modified. In some protected areas, the import of firewood by campers has been banned or discouraged in an effort to close this pathway for the entry of emerald ash borer (*Agrilus planipennis*) and other invasive forest pests and diseases. Managers of some protected waterways now require boats to be inspected or at least owner-certified to be clean and free of invasive species to prevent invasions by zebra mussels (*Dreissena polymorpha*), quagga mussels (*D. bugensis*), and other invasive aquatic plants and animals.

AMBITIOUS AND LARGE-SCALE STRATEGIES

In some situations only ambitious and large-scale invasive species control and eradication efforts will be adequate to achieve significant protected area conservation goals. Here, I echo the sentiments of Simberloff (2002, 2009) that we must aim high if we are to succeed in addressing invasive species threats and other anthropogenic threats to native biodiversity. Setting priorities remains crucial because the great costs associated with ambitious projects will make it possible to carry out only a fraction of even those with a high probability of success.

Despite their cost and complexity, successful eradication projects can yield large long-term savings in management costs and large benefits to native biodiversity. To date a variety of invasive vertebrates have been eradicated from islands and a few mainland protected areas, including mice, rats, rabbits, foxes, brushtail opossums, rock wallabies, feral burros, cats, cattle, dogs, goats, pigs, and sheep. With time and experience, vertebrate eradications have been carried out on larger and larger islands. Norway rats (*Rattus norvegicus*) were eradicated from 3,100-hectare Langara Island in Canada's Queen Charlotte Islands, and goats (*Capra hircus*) were declared eradicated from 58,465-hectare Santiago Island in the Galápagos (Simberloff 2009).

The impacts of eradication on target native species and communities were carefully monitored for some noteworthy projects, and a wide variety of benefits were documented, ranging from greatly increased abundance and cover of native animal and plant species to new detections of native species populations and even of native species thought to have gone extinct (Veitch and Clout 2002). In a number of cases, however, unexpected and undesirable effects were also documented, including ecological cascades triggered by sharp increases in the abundance of invasive plants after the eradication of invasive herbivores and the release of invasive prey species after eradication of predators (Bergstrom et al. 2009; Morrison 2008). Researchers and managers have recognized the need to monitor and be ready to react and adaptively manage the unexpected effects of eradications for more than a decade (Simberloff 2002).

On California's Santa Cruz Island, eradication of feral sheep (*Ovis aries*), carried out in stages in the 1980s and 1990s, had the desired effects of allowing many native plants, including endemic and rare species, to make a dramatic comeback on thousands of hectares that had been grazed bare and subjected to erosion by sheep and other introduced grazers (Morrison 2008; Veitch and Clout 2002). The recovery of native and nonnative vegetation on the 24,900-hectare island also had the unintended side effect of providing more food and cover for the feral pigs (*Sus scrofa*), which hindsight suggests should have been targeted for elimination first (Morrison 2008; Veitch and Clout 2002). By the late 1990s research inspired by the alarming decline of the endemic island fox (*Urocyon littoralis santacruzae*), a focal conservation target, revealed a hyperpredation scenario in which abundant feral pigs sustained a population of golden eagles (*Aquila chrysaetos*), which were new to the island, and allowed them to prey on the fox incidentally (Morrison 2008). Fox numbers dropped from about 1,500 individuals to fewer than 100 over several years. The Nature Conservancy and the National Park Service jointly developed and funded a plan to restore the fox population, which included a captive breeding program, removal of golden eagles, efforts to encourage the return of bald eagles (*Haliaeetus leucocephalus*), and eradication of feral pigs (*Sus domestica*). The pig eradication began in March 2005, and within 15 months pigs were no longer detectable. The program continued to gather data for 11 more months before announcing that eradication had been achieved in mid-2007. Captive-bred foxes have been returned to the wild, and their population has begun to increase (Morrison 2008).

Projects that tackle invasive species at a regional scale can limit the invader's impacts across the region and prevent their spread beyond it. The Nature Conservancy and partner organizations are undertaking such an effort on central Florida's Lake Wales Ridge, which supports one of the highest concentrations of endemic species in North America, including at least nineteen plant species and an endemic species of scrub jay (*Aphelocoma coerulescens*) (Gordon et al. 2004). The matrix system of pine savannas, interspersed by wetter communities and bisected by dry ridges, is highly vulnerable to invasion by introduced climbing ferns (*Lygodium japonicum* and *L. microphyllum*), which rapidly cover and shade out native vegetation and can ladder fires into forest and savanna canopies that are otherwise rarely exposed to fire (Gordon et al. 2004). Because climbing ferns can spread long distances by spores and rapidly increase in cover once established, the Central Florida Climbing Fern Strategy was launched in 2004 with the goal of creating a *Lygodium*-free region in the center of the state and preventing the spread of *L. microphyllum* northward. By 2008 *L. microphyllum* had been eliminated on more than 500 hectares, and project partners had plans to address its spread over 3 million hectares more (Doria Gordon, personal communication, 2008).

TRIAGE APPROACH TO PRIORITIZING STRATEGIES

Where more threats exist than can be addressed at once, managers may need to practice a form of triage (Hobbs et al. 2003), prioritizing resources where their use will be most efficient and effective. Unfortunately, some invasive species cannot be controlled with available methods, at least not at a scale necessary to protect target biological diversity. For example, despite decades of research and field trials, there is still no effective, affordable way to control large areas of cheatgrass (*Bromus tectorum*) and similar fire-promoting annual invasive grasses, which now dominate millions of hectares in the intermountain West. In these situations and others like them, the fact that there simply is no effective way to control the invader trumps other priority-setting factors. Low priority must be assigned to attempts to control them, even if they cause severe damage to focal conservation targets.

When faced with harmful invasive species such as cheatgrass that cannot be controlled, it is important to consider whether other actions can be taken to promote the survival and viability of target native species, populations, and communities and key ecological attributes and processes where the invaders are abundant. Managers and researchers have recently begun to explore this strategy, and already a few promising approaches have been suggested. For example, habitat refuges from predation may suffice to protect focal target native species where invasive predators cannot be controlled (Carter and Bright 2002; Stokes et al. 2004).

Prioritization and Triage in Hawaii Volcanoes National Park: A Case Study

Hawaii Volcanoes National Park is beset by a wide variety of invasive plants, animals, and pathogens. Among the most damaging are fire-promoting invasive grasses, particularly broomsedge (*Andropogon virginicus*) and beardgrass (*Schizachyruim condensatum*), which are native to the southeastern United States. Since the 1960s and 1970s, when these and other nonnative grasses invaded, the frequency of wildfires in the park increased by a factor of three, and the size of wildfires increased by a factor of sixty (Tunison et al. 2001). The native plants of the seasonally dry woodland dominated by ohi'a (*Metrosideros polymorpha*) are poorly adapted to fire, and grass-fueled wildfires have converted large areas of ohi'a woodland into savannas. Broomsedge, beardgrass, and other invasive fire-promoting grasses have also moved into coastal scrub and grassland vegetation at lower elevations in the park, particularly after feral goats were controlled. Fortunately, these coastal plant communities, and particularly some of the dominant species in them, are somewhat better adapted to fire. Despite several decades of efforts, managers and researchers from the park and various universities were unable to find effective methods for controlling the invasive grasses at a scale large enough to protect large areas of native vegetation (Tunison et al. 2001). They were also unable to find methods to prevent or contain damaging wildfires.

As a result, park scientists and managers were forced to reassess and revise their conservation goals. Now, in areas that formerly supported dry ohi'a woodlands, the goal is to identify and establish fire-tolerant native tree and shrub species that can persist or even spread into the nonnative grass-dominated savannas (Tunison et al. 2001). Managers have begun planting out and seeding more than fifteen native species in recently burned areas and plan to monitor their survival and reproduction over the next decade and more (Loh et al. 2004). In coastal scrub and grassland areas, the goal is to increase the abundance of fire-stimulated native grasses and to identify and establish native shrubs that can also tolerate frequent fires (Tunison et al. 2001). Although the long-term goal continues to be preservation of as much of the park's biodiversity as possible, for at least the mid-term the objective is to rehabilitate the system so that it both contains native plant species and remains resilient in the face of the new fire regime.

The Importance of Clear Goals and Priorities

Protected area managers should be aware that invasive species can cause severe and irreversible damage to native biological diversity and ecological integrity, even in remote wilderness areas. Actions to prevent new invasions and to eradicate, control, or abate the effects of invasive species can and should be taken to protect target biological diversity. Decades of such efforts and their resultant successes and failures have yielded two particularly important lessons. First, managers need to clearly articulate conservation goals and objectives in order to assess the relative importance of various threats, including invasive species, to the values of the protected area. This includes identifying the objects and processes of most concern, the target species, populations, communities, and processes that protected area managers should be managing for. Second, managers need to prioritize different invasive species threats and the strategies for addressing them. Priority setting should consider effects on native biodiversity, current and potential distribution and abundance, difficulty of control, and the biodiversity value of the areas invasive species occupy or may occupy.

Priorities must also be set among strategies to address invasive species threats, whether they be preventing new invasions; working to eradicate, contain, or control them; mitigating harmful impacts; or restoring native

---

BOX 10.1. LESSONS LEARNED FROM INVASIVE SPECIES MANAGEMENT

- Invasive species are a major threat to biological diversity and ecological integrity.
- It is necessary to identify and clearly articulate protected area goals and objectives before threats can be assessed.
- It is necessary to prioritize threats and the strategies to address them.
- In developing strategies for invasive species, careful consideration should be given to preventing new invasions, using a triage approach to allocate resources to strategies most likely to be effective, and deciding whether ambitious, large-scale strategies are necessary to achieve conservation goals.

biodiversity. Often, more threats exist than can be addressed at once, requiring managers to practice a form of triage that allocates resources to the strategies that will be most efficient and effective. Managers should carefully consider strategies to prevent new invasions because they may yield the greatest long-term benefits to biological diversity protection.

REFERENCES

Aplet, G. H. 1990. Alteration of earthworm community biomass by the alien *Myrica faya* in Hawaii. *Oecologia* 82:414–416.

APRS Implementation Team. 2000. *Alien plants ranking system*, version 5.1. Northern Prairie Wildlife Research Center, Jamestown, ND. Retrieved March 2, 2009 from www.npwrc.usgs.gov/resource/literatr/aprs/index.htm.

Bergstrom, D. M., A. Lucieer, K. Kiefer, J. Wasley, L. Belbin, T. K. Pedersen, and S. L. Chown. 2009. Indirect effects of invasive species removal devastate World Heritage Island. *Journal of Applied Ecology* 46:73–81.

Bøhn, T., P.-A. Amundsen, and A. Sparrow. 2008. Competitive exclusion after invasion? *Biological Invasions* 10:359–368.

Bradley, B. A., R. A. Houghton, J. F. Mustard, and S. P. Hamburg. 2006. Invasive grass reduces aboveground carbon stocks in shrublands of the western US. *Global Change Biology* 12:1815–1822.

Carter, S. P., and P. W. Bright. 2002. Habitat refuges as alternatives to predator control for the conservation of endangered Mauritian birds. Pp. 71–78 in C. R. Veitch and M. N. Clout, eds. *Turning the tide: The eradication of invasive species*. IUCN Species Survival Commission, Gland, Switzerland.

Croll, D. A., J. L. Maron, J. A. Estes, E. M. Danner, and G. V. Byrd. 2005. Introduced predators transform subarctic islands from grassland to tundra. *Science* 307:1959–1961.

De Poorter, M., S. Pagad, and M. I. Ullah. 2007. *Invasive alien species and protected areas: A scoping report*. The World Bank, Washington, DC. Retrieved March 2, 2009 from www.issg.org/Animal%20Imports%20Webpage/Electronic%20References/IASinPAs.pdf.

Gordon, D. R., J. L. Slapcinsky, B. Nelson, and S. Penfield. 2004. *Central Florida Lygodium strategy*. Paper presented at the annual meeting of the Association of Southeastern Biologists, Memphis, TN.

Hobbs, R. J., V. A. Cramer, and L. J. Kristjanson. 2003. What happens if we cannot fix it? Triage, palliative care and setting priorities in salinising landscapes. *Australian Journal of Botany* 51:647–653.

Layden, P. C., and M. J. Manfredo. 1994. *National park conditions: A survey of park superintendents*. National Parks and Conservation Association, Washington, DC.

Lesica, P., and D. Hanna. 2004. Indirect effects of biological control on plant diversity vary across sites in Montana grasslands. *Conservation Biology* 18:444–454.

Loh, R., A. Ainsworth, D. Benitez, S. McDaniel, M. Schultz, K. Smith, T. Tunison, and M. Valdya. 2004. *Broomsedge burn, Hawai'i Volcanoes National Park.* Burned Area Emergency Rehabilitation Final Accomplishment Report. U.S. National Park Service, Volcano, HI.

Martin, B., D. Hanna, N. Korb, and L. Frid. 2007. *Decision analysis of alternative invasive weed management strategies for three Montana landscapes.* The Nature Conservancy and ESSA Technologies Ltd., Helena, MT.

Meyers-Rice, B., and I. M. Tu. 2001. Weed management plan for Cosumnes River Preserve, Galt, California, 2001–2005. Retrieved October 13, 2009 from www.invasive.org/gist/products/plans/CRP-Plan.pdf.

Moody, M. E., and R. N. Mack. 1988. Controlling the spread of plant invasions: The importance of nascent foci. *Journal of Applied Ecology* 25:1009–1021.

Morrison, L. W. 2002. Long term impacts of an arthropod community invasion by the imported fire ant *Solenopsis invicta*. *Ecology* 83:2337–2345.

Morrison, S. A. 2008. Reducing risk and enhancing efficiency in non-native vertebrate removal efforts on islands: A 25 year multi-taxa retrospective from Santa Cruz Island, California. Pp. 398–409 in G. W. Witmer, W. C. Pitt, and K. A. Fagerstone, eds. *Managing vertebrate invasive species: Proceedings of an international symposium*. USDA/APHIS/WS, National Wildlife Research Center, Fort Collins, CO.

The Nature Conservancy. 2007. *Conservation action planning: Developing strategies, taking action and measuring success at any scale—Overview of basic practices*. The Nature Conservancy, Arlington, VA

Olmsted, F. L. 1865. Preliminary report upon the Yosemite and Big Tree Grove. Pp. 488–516 in V. P. Ranney, ed. *The papers of Frederick Law Olmsted V: The California frontier 1863–1865*. John Hopkins University Press, Baltimore, MD.

Parker, I. M., D. Simberloff, W. M. Lonsdale, K. Goodell, M. Wonham, P. M. Kareiva, M. H. Williamson, et al. 1999. Impact: Toward a framework for understanding the ecological effects of invaders. *Biological Invasions* 1:3–19.

Ricciardi, A., and J. Cohen. 2007. The invasiveness of an introduced species does not predict its impact. *Biological Invasions* 9:309–315.

Rinella, M. J., B. D. Maxwell, P. K. Fay, T. Weaver, and R. L. Sheley. 2009. Control effort exacerbates invasive-species problem. *Ecological Applications* 19:155–162.

Rodriquez, L. F. 2006. Can invasive species facilitate native species? Evidence of how, when, and why these impacts occur. *Biological Invasions* 8:927–939.

Shea, K., H. P. Possingham, W. W. Murdoch, and R. Roush. 2002. Active adaptive management in insect pest and weed control: Intervention with a plan for learning. *Ecological Applications* 12:927–936.

Simberloff, D. 2002. Today Tiritiri Matangi, tomorrow the world! Are we aiming too low in invasives control? Pp. 4–12 in C. R. Veitch and M. N. Clout, eds. *Turning the tide: The eradication of invasive species*. IUCN Species Survival Commission, Gland, Switzerland.

Simberloff, D. 2009. We can eliminate invasions or live with them: Successful management projects. *Biological Invasions* 11:149–157.

Simberloff, D., and B. Von Holle. 1999. Positive interactions of non-indigenous species: Invasional meltdown? *Biological Invasions* 1:21–32.

Stokes, V. L., R. P. Pech, P. B. Banks, and A. D. Arthur. 2004. Foraging behavior and habitat use by *Antechinus flavipes* and *Sminthopsis murina* (Marsupiala: Dasyuridae) in response to predation risk in eucalypt woodland. *Biological Conservation* 117:331–342.

Strayer, D. L., V. T. Eviner, J. M. Jeschke, and M. L. Pace. 2006. Understanding the long-term effects of species invasions. *Trends in Ecology and Evolution* 21:645–651.

Tunison, J. T., C. M. D'Antonio, and R. K. Loh. 2001. Fire and invasive plants in Hawai'i Volcanoes National Park. Pp. 122–131 in K. E. M. Galley and T. P. Wilson, eds. *Proceedings of the Invasive Species Workshop: The role of fire in the control and spread of invasive species*. Fire Conference 2000: The First National Congress on Fire Ecology, Prevention, and Management. Miscellaneous publication no. 11. Tall Timbers Research Station, Tallahassee, FL.

U.S. Department of Agriculture. 1999. Executive order 13112 of February 3, 1999: Invasive species. *Federal Register* 64(25):6183–6186. Retrieved October 13, 2009 from frwebgate.access.gpo.gov/cgi-bin/getdoc.cgi?dbname=1999_register&docid=99-3184-filed.pdf.

Veitch, C. R., and M. N. Clout, eds. 2002. *Turning the tide: The eradication of invasive species*. IUCN Species Survival Commission, Gland, Switzerland.

Wadsworth, R. A., Y. C. Collingham, S. G. Willis, B. Huntley, and P. E. Hulme. 2000. Simulating the spread and management of alien riparian weeds: Are they out of control? *Journal of Applied Ecology* 37:28–38.

Whisenant, S. G. 1990. Changing fire frequencies on Idaho's Snake River Plains: Ecological and management implications. Pp. 4–10 in E. D. McArthur, E. V. Romney, S. D. Smith, and P. T. Tueller, eds. *Proceedings: Symposium on cheatgrass invasion, shrub die-off, and other aspects of shrub biology and management*. General technical report INT-276. USDA Forest Service Intermountain Research Station, Las Vegas, NV.

Wilcove, D. S., D. Rothstein, J. Dubow, A. Phillips, and E. Losos. 1998. Quantifying threats to imperiled species in the United States. *BioScience* 48:607–615.

Wilson, E. O. 1992. *The diversity of life*. Belknap, Cambridge, MA.

Zavaléta, E. S., R. J. Hobbs, and H. A. Mooney. 2001. Viewing invasive species removal in a whole-ecosystem context. *Trends in Ecology and Evolution* 16:454–459.

Chapter 11

Responding to Climate Change: A Toolbox of Management Strategies

DAVID N. COLE, CONSTANCE I. MILLAR,
AND NATHAN L. STEPHENSON

The road to success is always under construction.

—*Lily Tomlin*

Climate change and its effects are writ large across the landscape and in the natural and cultural heritage of parks and wilderness. They always have been and always will be. The sculpted walls of Yosemite National Park and the jagged scenery of the Sierra Nevada wilderness would not be as spectacular if periods of glaciation had not been followed by periods of deglaciation. High biodiversity in forests of the Great Smoky Mountains reflects a legacy of climate change, migrating species, and isolated climatic refugia. Fossils unearthed at Dinosaur National Monument reflect a time when the climate was very different than it is today, as do ruins left by peoples who practiced agriculture in places in the American Southwest where food production is not possible today. Over eons, climate change has molded the diversity of life and landscape in areas now protected as parks and wilderness.

Contemporary climate change is quite different, however. For the first time, the pace and direction of climate change appear to be driven significantly by human activities (Intergovernmental Panel on Climate Change 2007). Contemporary climate change is playing out across landscapes already affected by other anthropogenic stressors—pollution, invasive species, altered disturbance regimes, and land fragmentation. Compared to landscapes with continuous habitat, fragmented landscapes severely diminish the ability of species to respond adaptively to rapid climate change.

Contemporary climate change therefore places much that humans value at risk. Biodiversity is threatened, as are precious park landscapes. The glaciers in Glacier National Park are projected to disappear by 2030 (Hall and Fagre 2003). Models suggest that climate change may make it impossible for Joshua trees (*Yucca brevifolia*) to persist in Joshua Tree National Park (Cole et al. 2005). Rising sea level threatens the freshwater wetlands of Everglades National Park. Increasingly severe drought, earlier snowmelt, reduced stream flows, larger and more intense fires, and widespread insect and disease infestations portend a future of diminished park and wilderness values, both social and ecological (Saunders et al. 2007).

The first step in responding to climate change is to clarify protected area goals and purposes. A prominent theme in this book has been the need to move beyond the singular traditional goal of sustaining naturalness and articulate diverse forward-looking goals more helpful in guiding when and how to intervene in ecosystems. Beyond this, there is an immediate need to begin planning for and responding to climate change. Some actions can be taken in the near term; others will bear fruit only in the future. Some actions can be implemented locally; others will be successful only when played out over large landscapes. In this chapter, we describe a toolbox of potential management responses to climate change. We begin with a general management framework, move to more specific near-term, local management actions, and conclude with longer-term, larger-scale approaches. Dealing with uncertainty—acknowledging it, taking action despite it, learning, and responding appropriately when surprises occur—is a central theme of the chapter.

Management Approaches in the Context of Climate Change

One of the actions protected area managers can take immediately is to incorporate climate change into existing management plans (Heller and Zavaleta 2009). Several frameworks for doing this have been proposed. Most involve articulating goals, identifying key ecosystem elements and processes, identifying indicators, setting baselines or limits of acceptable variation, assessing vulnerabilities and sensitivities, establishing monitoring programs, and identifying appropriate adaptive responses to climate change (Spittlehouse and Stewart 2003; Baron et al. 2008; Kareiva et al.

2008; Heller and Zavaleta 2009). For example, the National Park Service prepares a foundation plan as part of their general management plan. The Forest Service prepares a comprehensive evaluation report as part of their land management planning process. These documents focus on patterns and trends in environmental conditions, emerging stresses, and anticipated changes. Scenario planning exercises relevant to climate change (described in detail in Chapter 13) can be incorporated into strategic visions and frameworks that contribute to final plans.

The four management emphases explored in the second part of this book, autonomous nature, historical fidelity, ecological integrity, and resilience, represent a range of protected area goals with different management strategies. Where autonomous nature is the goal, the response to climate change is to not manipulate ecosystems. In contrast, to maintain a high degree of historical fidelity, intensive and frequent manipulation of ecosystems, much of it at localized and small scales, is often necessary. For example, Tuolumne Meadows in Yosemite National Park might be sustained as the largest subalpine meadow in the Sierra Nevada, offering the same stunning views in upcoming centuries as it has in the past. But this will probably require continual removal of lodgepole pine (*Pinus contorta*) seedlings and saplings that invade the meadow with increased frequency as climate changes, and possibly even irrigation to maintain species intolerant of drier conditions.

Ecological integrity and resilience are management goals that are in several ways intermediate between autonomous nature and historical fidelity. They imply active intervention to conserve elements of biodiversity, as managing for historical fidelity does, but are generally implemented at lower intensities and larger spatial scales than efforts to sustain historical fidelity. For example, with ecological integrity or resilience as a goal, lodgepole pine might be allowed to invade Tuolumne Meadows. Interventions might focus on ensuring that hydrologic regimes remain functional and the regional population viability of meadow species is not threatened.

If they are to intervene, managers must decide whether to be proactive, anticipate change and act in advance, or to respond only after disturbance or extreme events. Waiting to respond might reflect uncertainty regarding the need for action or what actions are likely to succeed. Waiting might reflect a precautionary approach and be preferred by risk-averse actors (Heller and Zavaleta 2009). Alternatively, waiting might reflect a decision that the best time for action, from a scientific or an organizational efficiency standpoint, is after disturbance. Regardless, what is important is to plan ahead, so responses can be quickly implemented when windows of opportunity open (Joyce et al. 2008). For instance, in the Rock Creek Butte Research Natural Area, California, the entire population of rare Brewer spruce (*Picea breweriana*) burned during a 2008 wildfire. Climate envelope modeling for this species suggests that there will no longer be favorable habitat for Brewer spruce in the future of California, and therefore restoration is not appropriate. However, after a fire is the best time to attempt regeneration and establish a refugium for Brewer spruce in its current native habitat.

Proactive, anticipatory responses use "current information about future climate, future environmental conditions and the future context" of protected area management "to begin making changes to policy and on-the-ground management now and when future windows of opportunity open" (Joyce et al. 2008: 3–40). The ability of climate science to make projections about future climate and resultant changes in biota has improved (Intergovernmental Panel on Climate Change 2007). However, as discussed in Chapter 4, the current state of the art suggests it is dangerous to commit to projections as accurate forecasts, especially at the scales relevant to park managers. Model outputs are better viewed as vehicles for organizing thinking, considering different scenarios, and gaining insight into a range of possible futures (Millar et al. 2007).

Adapting to Climate Change

Regardless of whether actions are taken reactively or proactively, the ultimate goal is to help ecosystems, ecological elements, and processes adapt to climate change and accomplish this by increasing the adaptive capacity of protected area policies and institutions. Many ways of adapting have been suggested, but most recommendations are still at the idea stage (Heller and Zavaleta 2009). Most are general rather than specific and actionable, based more on logical thinking than empirical data. This is particularly true for protected areas, where many specific recommendations (thinning forests, creating more diverse stand structures, realigning ecosystems, moving species) are potentially so intrusive and heavy-handed to seem anathema to traditional thinking about what is appropriate in wilderness and some parks. That is why this book emphasizes the need to rethink protected area goals and carefully evaluate whether proposed management actions advance or detract from goal achievement.

Some suggestions focus on ecological interventions, whereas others focus on societal issues, policy, and institutional change. Recommendations vary in the temporal and spatial context of application. In this section we first explore actions that can be taken in the near term by managers of individual parks and wildernesses. Then we turn to actions that take more time to implement and must be used at large spatial scales.

Near-Term Actions for Local Managers

We begin with near-term actions designed to protect ecosystems, buffering them from effects of climate change and helping them resist change. Ecosystems, their elements and processes, must respond to climate change by adapting in place, or species must migrate someplace else; otherwise they go extinct. Although resistance to change may seem a denial of future change, it is a defensible approach to uncertainty, particularly for highly valued attributes such as endangered or iconic species. Increasing resistance is also a means of buying time. A number of traditional stewardship actions will promote resistance. Managers can promote basic ecosystem functioning and mitigate threats to resources (Heller and Zavaleta 2009). Actions might include more aggressive management of adverse effects posed by invasive species, recreational use, livestock grazing, or water diversion. For instance, groundwater pumping for rural municipal use or for recreational developments such as ski areas can deplete mountain aquifers, causing springs, fens, and wetlands to dry up. Negotiated compromise and enhanced water conservation may reduce water demand.

Maintaining natural disturbance dynamics is another common recommendation (Taylor and Figgis 2007), but this option exposes the conflict between actions that create resistance to change and those that promote adaptive capacity or resilience. For example, climate change is expected to increase fire frequency and intensity, a change already being observed (Westerling et al. 2006). Suppressing unusually severe fires, particularly those that threaten highly valued ecosystem elements, such as old-growth forests, is a means of resisting changes wrought by climate, but this is often not possible. Moreover, doing so ignores the fact that many biotic elements need fire for their persistence. Intervening in fire processes interferes with an important mechanism by which vegetation adjusts to new climatic conditions. As Noss (2001: 585) suggests, "a mixed strategy in which managers let many natural fires burn, protect (to the extent possible) old growth from stand-replacing fires, and manage other stands by prescribed burning and understory thinning to reduce the risk of high-intensity fire, may be the optimal approach." Indeed the very concept of natural disturbance is evolving as changing climate brings fire and insect outbreaks to places that have not experienced them for centuries, at larger spatial scales and at different times of the year than in the past.

Over paleohistorical time scales, climate has acted as a driver of biotic change, with much of that change occurring synchronously across the landscape (Betancourt et al. 2004). At decadal and centennial scales, for example, windstorms in the eastern United States and drought in the West have at times synchronized forest composition and structure across the landscape, possibly making them more vulnerable to climatic shifts. A recent example may be the extensive dieback in some forests as a result of drought (Breshears et al. 2005). This has led some to suggest that vulnerability to climate change might be reduced by deliberate reduction of landscape synchrony (Millar and Woolfenden 1999; Betancourt et al. 2004). Actions that promote diverse age classes, species mixes, and landscape structural and genetic diversity reduce landscape synchrony. This can be done most effectively at early successional stages, when ecological trajectories are influenced more by present and future climatic conditions than by those of the past (Millar et al. 2007).

Peters and Darling (1985) suggest it may sometimes be necessary to undertake heroic rescue efforts, for example by irrigating sensitive species. This might be one response to the potential loss of Joshua trees (*Yucca brevifolia*) in Joshua Tree National Park. Other examples include using attractants to lure songbirds to continue to use specific meadows, providing winter forage during harsh years for endangered species, or providing supplemental forage to encourage dispersal along planned corridors. Such projects illustrate the degree to which enhancing resistance in the face of directional climate change is akin to paddling upstream (Millar et al. 2007). They require ever-increasing effort, of a nature so intensive as to often be deemed undesirable, particularly in wilderness. Moreover, if conditions change enough, all but the most localized and intensive efforts may be futile. Ecosystems are likely to cross thresholds and be lost, perhaps with catastrophic consequences (Harris et al. 2006). Considering this, Millar et al. (2007) conclude that resistance options are best applied for the short term, to buy time, and where values at risk are high or sensitivity to climate change is low.

Complementary to enhancing resistance is facilitating the ability of ecosystems to adapt to climate change. Some options involve increasing capacity to adapt in place; others involve facilitating migration. Where ecosystems have been significantly disturbed, restoration treatments are often prescribed. Rather than restoring ecosystems to historical predisturbance conditions, managers increase adaptive capacity by realigning ecosystems with current and future conditions. Examples include restoration of fire regimes altered by fire suppression or of stream or meadow hydrologic relations altered by water impoundments, diversions, or livestock grazing. In situations where there is substantial confidence in predictions about future conditions, management interventions can be narrowly targeted. Fires may need to burn more frequently or during different seasons than in the recent past. Managers may want to encourage compositional shifts toward more drought-tolerant species where meadows have been altered by grazing or water diversion. More commonly, where the future is quite unpredictable, it might be best to realign ecosystems with a range of possible future conditions, for example using a broad species mix rather than a mix targeted to past conditions.

Traditional genetic management guidelines aim to avoid contamination of populations with ill-adapted genotypes by establishing small seed zones and dictating that seeds (or other plant materials) used for restoration come from small, local zones. This restriction on genetic diversity made sense under the assumption that environments and climates were stable. But relaxing genetic guidelines, for both plants and animals, using germplasm from a wider geographic area and diversity of populations, makes more sense given climatic change (Ledig and Kitzmiller 1992). Managers might use predictions that future climates will be warmer to emphasize germplasm from warmer (often downhill) populations. Alternatively, they might hedge their bets by enlarging seed zones in all directions (Joyce et al. 2008). Best genetic management practices also emphasize equalizing germplasm contributions and enhancing population sizes by maximizing the number of parents and striving for equal amounts of plant material from each parent. These guidelines become especially important in an uncertain future.

Species unable to adapt in place must move. To migrate successfully, species need viable source populations and habitats, appropriate destinations, and a way to get from source to destination (Taylor and Figgis 2007). Managers can assist by identifying and conserving refugia, environments that are buffered against climate change and other disturbances. Refugia provide places where a species might persist even if unfavorable conditions cause it to disappear elsewhere. Refugia can be important sources and destinations. Past climatic refugia can be identified. For example, mountainous landscapes have highly heterogeneous environments, with diverse microclimates, some of which (particularly cool and mesic sites) can act as refugia. Later in this chapter we describe a case in which managers are reorienting goals to optimize an area's value as a climatic refugium. Once identified, refugia and the populations they harbor become high priorities for protection.

Species movement can be facilitated by allowing migration. As rates of species movement increase in response to change, traditional notions of a species range will be challenged, and the line between native and non-native species will blur, as will the definition of an invasive species (Millar and Brubaker 2006). Managers will need to decide whether invasions are beneficial or adverse and manage accordingly. Managers may even manipulate environments to encourage migration, thinning forests, for example, to encourage species that establish and grow better with more light and less competition.

More controversial is assisted migration, actively helping propagules or individuals move to new habitats where they are presumably better adapted. Assisted migration has sparked much interest and debate. Translocations are often unsuccessful, and unanticipated consequences can result from introducing new species into extant communities (Heller and Zavaleta 2009). Hoegh-Guldberg et al. (2008) propose a framework for deciding where assisted migration seems necessary and feasible and where other options are preferred. McLachlan et al. (2006) argue that there is an urgent need for debate and policy development regarding assisted migration. To start the process, these authors identify major policy choices, articulate implications of each choice, and provide a research agenda to inform debate and decision.

The final option is ex situ protection of the species most threatened by climate change, where this is the only option short of extinction. Species could be preserved in zoos and botanical gardens in the hopes that a time will come when the effects of climate change can be reversed and species can be returned to their native landscapes (Noss 2001).

Longer-Term, Larger-Scale Actions

As noted earlier, many near-term actions that local protected area managers might take are likely to be controversial, involving intensive and manipulative actions, often undertaken under conditions of high uncertainty. Perhaps less controversial, probably more important, but more difficult to implement are actions that must be taken at large spatial scales and played out over long timeframes. Chapters 12 and 13, on conservation at large spatial scales and building more adaptability into planning, cover some of this same material. Consequently, discussions here are brief and more suggestive than definitive.

One challenge to protected area management is that species move but protected area boundaries are fixed. This has long been a problem where protected areas are small and when animal populations migrate. Climate change will exacerbate this problem. The obvious response is to plan for conservation at much larger scales, enlarging the effective size of preserves and reducing limitations associated with their finite locations in space. There are several ways regional and landscape planning can improve adaptation to climate change.

Because species have to move, an obvious goal is to promote landscape connectivity (Noss 2001). Plants (and many animals) move by dying out in places that are no longer hospitable and colonizing newly habitable places. These migrations are inhibited to the degree that landscapes are fragmented and encouraged by the connectivity of the landscape. Landscapes with continuous habitat and few physical or biotic impediments to migration provide good connectivity (Millar and Brubaker 2006). Fragmentation can be avoided (e.g., by limiting roads), and connectivity can be promoted (e.g., by providing underpasses where there are roads) within large protected areas. But if the biotic shifts associated with climate change are substantial, connectivity must extend beyond individual protected areas to networks of protected areas and to the lands that constitute the matrix in which protected areas are situated. As discussed further in Chapter 12, this has led to proposals that core protected areas be linked by corridors to facilitate movement and exchange (see Noss et al. 1999, for example) and proposals for continental-scale corridors and linkages.

Connectivity can be promoted in many ways at various scales. Examples include situating trails in parks where they will have minimal connectivity impacts on small rodents and lagomorphs and building campsites where they will not interrupt migration routes of large mammals. Similarly, vegetation can be managed to maintain unimpaired wildlife movement. For example, prescribed fire has been used in the Sierra Nevada to reduce hiding cover for mountain lions (*Felis concolor*), the presence of which inhibited movement of Sierra bighorn sheep (*Ovis canadensis sierrae*) upslope to cooler, moister summer habitat.

A related action involves managing the matrix, the lands around and between protected areas that are not designated for protection but increasingly influence the integrity of parks and wilderness. The matrix can be a source of threats, especially if some lands harbor invasive species or generate pollution. The matrix can be fragmented and pose significant barriers to connectivity. Ideally, conservation in the matrix would "soften" land use by encouraging less damaging practices, such as practicing low-intensity forestry rather than clearcutting (Heller and Zavaleta 2009). Barriers to this strategy are substantial because many landowners and mangers have goals quite different from those of nearby protected areas, often emphasizing development and commodity extraction. Matrix lands are often owned by other government agencies or are private lands. Collaboration and identification of shared goals are needed, as are innovative mechanisms such as conservation easements and conservation futures (Hannah et al. 2002).

Regional planning is critical to promoting diversity and redundancy in management strategies as a means of managing risk associated with the uncertainties associated with climate change, its biotic effects, and the effectiveness of responses to change. Diversification is a means of hedging bets (Hummel et al. 2008), not putting all one's eggs in the same basket (Millar et al. 2007), and increasing adaptive capacity. Returning to the goals of Part II, the likelihood of optimizing all protected area values is increased if varied goals and strategies are used in different protected areas or different parts of large protected areas. This diversified approach increases the likelihood that if one strategy fails, a different strategy might succeed elsewhere. The complement of diversity, redundancy, is equally important. Diverse management strategies should be replicated across diverse environments. Again, if they fail in one place, they might succeed elsewhere. Ensuring planned, purposeful diversity and redundancy entails regional-scale planning.

The challenge of coordinated planning beyond the boundaries of protected areas and among networks of protected areas suggests the need for innovation. New goals must be articulated. These goals should be diverse, rather than monolithic, if they are to capture the array of protected area purposes and values. In many situations, goals may focus more on function or process than on composition and structure (Millar and Brubaker 2006). They may focus on species persistence in large geographic areas, relaxing expectations that current species ranges will remain constant or that population abundances, distributions, and species composition will remain stable (Millar et al. 2007). Seastedt et al. (2008) note that it may be better to manipulate mechanisms that enhance desirable system components than to remove or suppress undesirable components. A broader range of ecosystem types may be considered desirable (Heller and Zavaleta 2009). Rather than target a single desired future condition, goals might shift to avoiding a range of undesirable conditions (Joyce et al. 2008).

Managers must anticipate, confront, and incorporate uncertainty and the likelihood of surprise into planning and management. They must accept that climate change is likely to push populations beyond thresholds of mortality (Millar et al. 2007), sometimes to highly degraded states, and often to conditions for which there is no analog today or in the past (Harris et al. 2006). Where managers can control ecological processes to a substantial degree, adaptive management is an effective means of dealing with uncertainty. Discussed at length in Chapter 13, adaptive management involves learning by doing. Managers design actions that test uncertainties, monitor results, learn, and adjust practices accordingly.

Where uncertainty is high and managerial control is limited, scenario planning is a useful tool (Peterson et al. 2003). Scenario planning involves articulating and exploring a wide set of alternative futures, each of which is plausible but uncertain. Scenarios provide a mechanism for anticipating and working through conflicts between goals. Contingency plans can be developed for observable undesirable trends that are likely to continue and for catastrophic events with a low probability of occurrence (Baron et al. 2008).

Conflict, trade-offs, uncertainty, and limited resources suggest the need to prioritize and practice triage. Goals must be prioritized, as must the ecological elements and processes managers seek to sustain. Regarding the protection of selected elements or prioritizing of specific management situations, a triage approach might be helpful (Hobbs et al. 2003). In a resource context, triage involves systematically sorting different situations on the basis of urgency, sensitivity, and capacity of available resources to achieve desired outcomes (Millar et al. 2007). Categories range from situations with high urgency, adequate resources, and a proposed treatment that is likely to be effective (treat immediately) to situations that are untreatable, regardless of urgency, because of inadequate resources, catastrophic degradation, or no viable treatment.

Most of the actions discussed here—from planning for diversity and redundancy to prioritization and triage—are best done at scales that extend far beyond the boundaries of individual protected areas. This raises the need for institutional change to improve regional coordination through increased interagency cooperation and cooperation between different field units in the same agency. As discussed in Chapter 12, it is imperative to develop the institutional capacity to produce regional visions and strategies so that local decisions made by managers of individual protected areas add to the diversity, redundancy, and capacity of regional systems. This will be challenging given the decentralized decision-making tradition of public land agencies and traditional lack of cooperation between agencies and between public and private lands.

Equally important are efforts to change policies and institutions to enhance flexibility and the capacity to adapt through learning. To confront uncertainty and avoid paralysis, appropriate risk taking must be encouraged. "Safe to fail" strategies intend to succeed but recognize the potential for failure. Punishing managers who prudently accept risk but ultimately fail will make all managers so risk averse that proactive actions that are needed will never be taken (Baron et al. 2008). Although never desired, failure opens the door to learning and whittling away at uncertainty. But to learn from failure, we must monitor the effects of actions, with lessons learned incorporated into future plans. Managers must develop the capacity to reassess conditions frequently and be willing to change course as conditions change. As discussed in Chapter 13, this will require institutions and policies that emphasize flexibility rather than highly structured decision making (Millar et al. 2007).

Devil's Postpile National Monument as Climatic Refugium: A Case Study

Despite the fact that effects of climate change are already apparent in many protected areas, few have addressed climate change in a substantial manner. Devil's Postpile National Monument in California is one unit attempting to assess implications of climate change for their stewardship program and even revisit park purposes and goals.

Devil's Postpile is a small park unit (325 hectares) located at about 2,300 meters elevation, close to the headwaters of the San Joaquin River in the Sierra Nevada. It is adjacent to Forest Service land, much of it wilderness, not far from Yosemite National Park and the major destination ski resort Mammoth Mountain. The monument was created to protect one of the world's finest examples of highly symmetrical columnar basalt (Figure 11.1) and is known for its fine mountain scenery. Despite its small size, the monument is highly diverse, with more than 400 plant species, 100 bird species, 12 species of bats, and 35 mammal species. Some of the reasons for this biodiversity include the prevalence of wetlands in the monument (8.5 percent of the park) and its location at a low point in the Sierra Divide, where three bioregions converge (Central, Southern, and Eastern Sierra).

Research suggests that the monument might provide an important refugium for many species during a period of climate change because of its

FIGURE 11.1. Devil's Postpile National Monument was established to protect a unique geological resource, perhaps the world's finest example of highly symmetrical columnar basalt. In a future of rapid climate change, it may help sustain regional biodiversity by being an important climatic refugium. (Photo by the National Park Service)

unique topographic position. Located in a deep canyon, running north to south, the monument is subject to substantial pooling of cold air. Projections suggest that this cold air pool might warm more slowly than the surrounding landscape (if it warms at all), buffering the park from predicted warming and drying and allowing the park's wetlands to survive longer than might be the case elsewhere (Lundquist and Cayan 2007). Managers of the monument are considering a reorientation of their goals and purposes to optimize the monument's potential as a climatic refugium, an area that is less affected by climate change than its surroundings.

Monument managers are developing a strategy to build on current programs to increase their capacity to plan for and respond to climate change. They are attempting to obtain additional resources and staff to study, monitor, and manage issues such as climate, past and present, and its effects on native species, invasive species, pest outbreaks, and the need for restoration. In recognition of the need to work at spatial scales much larger than the monument itself, collaborative relationships are being developed with other land management agencies and private entities that own or manage surrounding lands. Interagency workshops have been held, in an effort to manage the Eastern Sierra as more of a unified system. This is extremely important, because the monument functions as a critical migration corridor, connecting lower and higher elevations, and its wetlands are sustained by waters that originate on national forest land and that can be adversely affected by development on neighboring private lands.

Moving Forward in the Face of Uncertainty and Change

Climate change is likely to be the defining issue facing managers of parks and wilderness in the twenty-first century. Ecological changes are already occurring, and there is a rising sense of urgency. But what exactly should be done? In this chapter, we have described a variety of actions that local managers can consider in the near term. Many of these actions are risky. Uncertainty is high, and some actions are so intensive and manipulative as to seem inappropriate on lands that are supposed to be wild and uncontrolled. Less controversial is the urgent need to change policies and institutions to make them more flexible and adaptive and thereby increase the resilience of ecosystems to climate change.

Millar et al. (2007: 2146) describe the institutional milieu that is necessary, one that embraces "strategic flexibility, characterized by risk-taking (including decisions of no action), capacity to reassess conditions frequently, and willingness to change course as conditions change." Park and wilderness goals should be more diverse, both to be responsive to varied park purposes and to reduce risk. Such goals will probably need to be more flexible, allowing for a broader array of desirable futures, more tolerant of biotic changes resulting from human activity (Welch 2005), and couched more in terms of maintaining regional biodiversity and ecosystem function than maintaining contemporary biotic community composition and structure.

Successful regional planning and collaboration beyond protected area boundaries will be critical to responding effectively to climate change. Parks have always been too small to manage in isolation; climate change makes them that much smaller. What are urgently needed are institutions and policies that enable planned diversity and redundancy in goals, strategies, and management practices across large landscapes and promote connected landscapes to ease the movement of species.

BOX 11.1. RESPONDING TO CLIMATE CHANGE

- Managing in the face of climate change requires a toolbox of approaches, including short-term and long-term strategies that focus on ecosystem resistance and resilience and help ecosystems adapt to change.
- Dealing with uncertainty is a fundamental challenge best dealt with through increased flexibility and adaptability, as well as carefully planned diversity and redundancy at multiple scales.
- Near-term actions managers of individual protected areas might consider include the following:
    Mitigating threats to resources
    Maintaining natural disturbance dynamics
    Reducing landscape synchrony
    Making aggressive but thoughtfully prioritized efforts to rescue highly sensitive species
    Realigning conditions with current, expected, or a range of possible future conditions
    Relaxing genetic guidelines where risk is low and adaptive management can be implemented
    Conserving refugia
    Allowing or actively assisting migration
    Cautiously considering the use of nonnative species where they are the best option for maintaining critical ecosystem functions
    Protecting highly endangered species ex situ
- Longer-term, larger-scale actions include the following:
    Promoting landscape connectivity
    Managing the matrix
    Promoting diversity and redundancy
    Articulating new goals
    Incorporating uncertainty and the likelihood of surprise into planning and management
    Prioritizing and practicing triage
    Increasing interagency cooperation
    Increasing flexibility and the capacity to adapt through learning

REFERENCES

Baron, J. S., C. D. Allen, E. Fleishman, L. Gunderson, D. McKenzie, L. Meyerson, J. Oropeza, and N. Stephenson. 2008. National parks. Pp. 4-1–4-68 in S. H. Julius and J. M. West, ed. *Preliminary review of adaptation options for climate-sensitive ecosystems and resources. A report by the U.S. Climate Change Science Program and the Subcommittee on Global Change Research*. U.S. Environmental Protection Agency, Washington, DC.

Betancourt, J., D. Breshears, and P. Mulholland. 2004. *Ecological impacts of climate change*. Report from a NEON science workshop, August 24–25, 2004, Tucson, AZ. American Institute of Biological Sciences, Washington, DC.

Breshears, D. D., N. S. Cobb, P. M. Rich, K. P. Price, C. D. Allen, R. G. Balice, W. H. Romme, J. H. Kastens, M. L. Floyd, and J. Belnap. 2005. Regional vegetation die-off in response to global-change–type drought. *Proceedings of the National Academy of Sciences (USA)* 102:15144–15148.

Cole, K. L., K. Ironside, P. Duffy, and S. Arundel. 2005. *Transient dynamics of vegetation response to past and future climatic changes in the southwestern United States*. Retrieved April 2, 2008 from www.climatescience.gov/workshop2005/posters/P-EC4.2_Cole.pdf.

Hall, M., and D. Fagre. 2003. Modeled climate-induced glacier change in Glacier National Park, 1850–2100. *BioScience* 53:131–139.

Hannah, L., G. F. Midgley, and D. Millar. 2002. Climate change–integrated conservation strategies. *Global Ecology and Biogeography* 11:485–495.

Harris, J. A., R. J. Hobbs, E. Higgs, and J. Aronson. 2006. Ecological restoration and global climate change. *Restoration Ecology* 14:170–176.

Heller, N. E., and E. S. Zavaleta. 2009. Biodiversity management in the face of climate change: A review of 22 years of recommendations. *Biological Conservation* 142:14–32.

Hobbs, R. J., V. A. Cramer, and L. J. Kristjanson. 2003. What happens if we cannot fix it? Triage, palliative care and setting priorities in salinising landscapes. *Australian Journal of Botany* 51:647–653.

Hoegh-Guldberg, O., L. Hughes, S. McIntyre, D. B. Lindenmayer, C. Parmesan, H. P. Possingham, and C. D. Thomas. 2008. Assisted colonization and rapid climate change. *Science* 321:345–346.

Hummel, S., G. H. Donovan, T. A. Spies, and M. A. Hemstrom. 2008. Conserving biodiversity using risk management: Hoax or hope? *Frontiers in Ecology and the Environment* 7:103–109.

Intergovernmental Panel on Climate Change. 2007. *Climate change 2007: Synthesis report*. Retrieved February 25, 2009 from www.ipcc.ch/pdf/assessment-report/ar4/syr/ar4_syr.pdf.

Joyce, L. A., G. M. Blate, J. S. Littell, S. G. McNulty, C. I. Millar, S. C. Moser, R. P. Neilson, K. O'Halloran, and D. L. Peterson. 2008. National forests. Pp. 3-1–3-127 in S. H. Julius and J. M. West, eds. *Preliminary review of adaptation options for climate-sensitive ecosystems and resources: A report by the U.S. Climate Change Science Program and the Subcommittee on Global Change Research*. U.S. Environmental Protection Agency, Washington, DC.

Kareiva, P., C. Enquist, A. Johnson, S. H. Julius, J. Lawler, B. Petersen, L. Pitelka, R. Shaw, and J. M. West. 2008. Synthesis and conclusions. Pp. 9-1–9-66 in S. H. Julius and J. M. West, eds. *Preliminary review of adaptation options for cli-*

mate-sensitive ecosystems and resources: A report by the U.S. Climate Change Science Program and the Subcommittee on Global Change Research. U.S. Environmental Protection Agency, Washington, DC.

Ledig, F. T., and J. H. Kitzmiller. 1992. Genetic strategies for reforestation in the face of global climate change. *Forest Ecology and Management* 50:153–169.

Lundquist, J. D., and D. R. Cayan. 2007. Surface temperature patterns in complex terrain: Daily variations and long-term change in the central Sierra Nevada, California. *Journal of Geophysical Research* 112, D11124, doi:10.1029/2006JD007561.

McLachlan, J. S., J. L. Hellmann, and M. W. Schwartz. 2006. A framework for debate of assisted migration in an era of climate change. *Conservation Biology* 21:297–302.

Millar, C. I., and L. B. Brubaker. 2006. Climate change and paleoecology: New contexts for restoration ecology. Pp. 315–340 in M. Palmer, D. Falk, and J. Zedler, eds. *Foundations of restoration ecology: The science and practice of ecological restoration*. Island Press, Washington, DC.

Millar, C. I., N. L. Stephenson, and S. L. Stephens. 2007. Climate change and forests of the future: Managing in the face of uncertainty. *Ecological Applications* 17:2145–2151.

Millar, C. I., and W. B. Woolfenden. 1999. The role of climate change in interpreting historic variability. *Ecological Applications* 9:1207–1216.

Noss, R. F. 2001. Beyond Kyoto: Forest management in a time of rapid climate change. *Conservation Biology* 15:587–590.

Noss, R. F., J. R. Strittholt, K. Vance-Borland, C. Carroll, and P. Frost. 1999. A conservation plan for the Klamath–Siskiyou ecoregion. *Natural Areas Journal* 19:392–411.

Peters, R. L., and J. D. S. Darling. 1985. The greenhouse-effect and nature preserves. *BioScience* 35:707–717.

Peterson, G. D., G. S. Cumming, and S. R. Carpenter. 2003. Scenario planning: A tool for conservation in an uncertain world. *Conservation Biology* 17:358–366.

Saunders, S., T. Easley, J. A. Logan, and T. Spencer. 2007. Losing ground: Western national parks endangered by climate disruption. *George Wright Forum* 24:41–81.

Seastedt, T. R., R. J. Hobbs, and K. N. Suding. 2008. Management of novel ecosystems: Are novel approaches required? *Frontiers in Ecology and the Environment* 6:547–553.

Spittlehouse, D. L., and R. B. Stewart. 2003. Adaptation to climate change in forest management. *British Columbia Journal of Ecosystems and Management* 4:1–11.

Taylor, M., and P. Figgis. 2007. Protected areas: Buffering nature against climate change—Overview and recommendations. Pp. 1–12 in M. Taylor and P. Figgis, eds. *Protected areas: Buffering nature against climate change*. Proceedings of a WWF and IUCN World Commission on Protected Areas symposium, June 18–19, 2007, Canberra. World Wildlife Fund–Australia, Sydney.

Welch, D. 2005. What should protected areas managers do in the face of climate change? *The George Wright Forum* 22:75–93.

Westerling, A. L., H. G. Hidalgo, D. R. Cayan, and T. W. Swetnam. 2006. Warming and earlier spring increase western U.S. forest wildfire activity. *Science* 313:940–943.

Chapter 12

Conservation at Large Scales: Systems of Protected Areas and Protected Areas in the Matrix

PETER S. WHITE, LAURIE YUNG, DAVID N. COLE, AND RICHARD J. HOBBS

Inside the museums, infinity goes up on trial.

—*Bob Dylan*

When the first national parks were created, the implicit assumption was that they were like museums in the sense that they would protect, in perpetuity, the nature they contained—the landscapes, species, and ecosystems present historically. In this chapter, we argue that national parks and other protected areas must be managed as a distributed network of protected areas within larger bioregional landscapes, not only to retain particular sets of ecosystems or species in place but also to conserve the ability of ecosystems and species to shift spatially and adapt to changing environments. In this sense, infinity (in the language of protected area policy this becomes "in perpetuity" and "for future generations") is indeed on trial.

The concern that protected areas cannot fully accomplish conservation goals by themselves is based on four observations. First, the largest protected areas, even those 100,000 to 1 million hectares in size, are not large enough to circumscribe all the ecological processes that are important to the long-term sustainability of historic ecosystems. Second, protected area boundaries are permeable to many human influences, leaving a park's biodiversity vulnerable to the impacts of air pollution, water pollution, and invasive species. Third, connectivity between protected areas varies widely, such that gene flow and the ability of species to recolonize after local extinctions are compromised. Finally, in an era of climate change conservation success will depend on the interaction of protected areas and their surroundings and on the number, connectedness, and spatial arrangement of protected areas.

Earlier chapters in this book describe how the varied goals of parks and wilderness areas, once thought to be comfortably subsumed within the singular concept of naturalness, can be met only through diverse management approaches. A last reason for large-scale planning is that pursuing this diversity, although it is sometimes applied at the scale of the individual protected area, must be planned at much larger spatial scales. Whether for biological or management reasons, resilience increases when planning is conducted at landscape and regional scales (Taylor and Figgis 2007).

Although species distributions and ecosystems will change (as they always have), the boundaries of protected areas are fixed in space as legal management units. "Thinking outside the box" is literal in this case: We must think beyond the boundaries of individual protected areas to maximize conservation success. In this chapter we review the scientific basis for conservation at large spatial scales. We discuss three rationales for large-scale conservation planning: coordinating between separate protected areas; planning the spatial configuration of protected areas into networks, nodes, and corridors; and managing protected areas and the surrounding matrix. We explore barriers to large-scale conservation and describe necessary institutional and policy change. We conclude with a case study from the Greater Yellowstone Ecosystem.

The Scientific Basis for Large-Scale Conservation Planning

At the large scales needed for ecosystem management, human use areas are inevitably part of an overall conservation design that includes core protected areas. Ideas about conservation at larger scales can be traced in North American conservation at least to 1932 (Grumbine 1994). In that year, the Ecological Society of America's Committee for the Study of Plant and Animal Communities recognized that a comprehensive system of sanctuaries should represent a wide range of ecosystems, be managed for natural disturbances, and use a core and buffer approach (Shelford 1933). Since that report, four scientific areas have shaped large-scale conservation planning: island biogeography, metapopulation biology, disturbance ecology, and landscape ecology. We review these research areas and summarize principles for large-scale conservation based on them.

Island Biogeography

The theory of island biogeography (MacArthur and Wilson 1967) predicts that the species richness of islands reaches an equilibrium determined by immigration rates, which decrease as island isolation increases, and extinction rates, which increase as island size decreases. When applied to habitat fragmentation, the theory predicts that species richness will decline because of the increased isolation and decreased size of remaining habitat. Diamond (1975) uses the theory of island biogeography to argue that protected areas should be as large and as connected as possible, and narrow shapes should be avoided to minimize edge effects.

How large must protected areas be? This question is most often answered by determination of the area needed for the most area-demanding species, among which large mammalian predators are often the most important to consider because of their effects on overall ecosystem structure (Terborgh et al. 1999). Large mammalian predators need 1 million hectares for a minimum viable population size of 1,000 (Schoenwald-Cox 1983). Newmark (1987) showed that postpark extinctions in the western United States were minimal only if park size was greater than 1 million hectares. Given that the larger national parks in the lower forty-eight states average about 160,000 hectares, it is clear that cooperative management across protected area boundaries will be necessary to ensure that species needing large areas persist. Larger-scale conservation is important for other reasons as well, including seasonal migrations, water quality, and disturbance-driven dynamics. Even in Alaska, where the large wilderness areas average 1.7 million hectares, managers must address scales larger than individual units because of seasonal migrations and climate change. The question that emerges from this work is, "How can management in bioregional landscapes increase the effective size of protected areas?"

Metapopulations

Metapopulations are sets of subpopulations that are linked over larger scales by the dispersal of individuals (Levins 1969). The most basic question addressed by metapopulation biology is how the subpopulations contribute to regional persistence despite local fluctuation in abundance and even transient local extinction. As with the theory of island biogeography, size and isolation (of the subpopulations) govern the answer to this question.

Subpopulations in a metapopulation vary in size. Large subpopulations in high-quality habitat are often source populations in which reproduction exceeds mortality, so that the subpopulation can act as a source of dispersing individuals to other subpopulations. Smaller subpopulations in smaller patches and poorer-quality habitat are sink populations in which mortality exceeds reproduction. High-quality hotspots of reproduction are critical to metapopulation persistence, but even sink populations can be important when they allow stepping stone dispersal and thus the reestablishment of subpopulations after local extinctions (Foppen et al. 2000). As with island biogeography, both large habitat patches and connectivity are important to species persistence.

Disturbance

Research on the ecology of natural disturbances has further underscored the need for large-scale conservation planning (see White and Jentsch 2001). Pickett and Thompson (1978) argue that managing for disturbance and succession requires areas larger than the size of an individual disturbance event so that all species and age states (and therefore the continuation of the dynamic pattern) will persist. They call the area needed to sustain dynamics the minimum dynamic area. Disturbance ecology influenced the development of the concept of the historical range of variability as a guide to ecosystem management (see Chapters 2 and 8) by showing that ecosystems exhibit variation in space and through time because of disturbance.

A concept that developed in parallel with the historical range of variability was the idea that ecosystems could exhibit dynamic equilibrium under a given disturbance regime. The essence of this idea is that individual locations are periodically disturbed and then undergo succession, but ecosystem attributes (e.g., biomass across a landscape) averaged across many locations of different ages can be nearly constant. Spatial scale is a critical condition for dynamic equilibrium (Turner et al. 1993). Two extreme forms of dynamic equilibrium were defined: quantitative and qualitative equilibrium. Quantitative equilibrium is a strict equilibrium with no variance in the parameter of interest (e.g., the percentage of the land disturbed each year). Qualitative or persistence equilibrium is equilibrium with low to high variance (reviewed in White et al. 1999). For example, the 1988 fires in Yellowstone National Park affected a large percentage of the landscape, but the park as a whole meets the conditions for dynamic equilibrium, with high variance under the fire regime of the last three centuries (Turner et al. 1993). The area needed for disturbance dynamics (the minimum dynamic area to allow persistence of species, ecosystem variation, and disturbances) depends on the details of the disturbance regime, but studies suggest that the total area must be 50–100 times the size of disturbance patches (Shugart 1984; Turner et al. 1993).

The idea that large landscapes could simultaneously be characterized by local dynamics and larger-scale persistence was attractive in conservation management because it suggested that ecosystems could be managed at a scale appropriate to this dynamic stability. The scale of such dynamics can exceed the size of protected areas, but there is a further reason for managing at larger scales: Natural disturbances do not respect the political boundaries of protected areas (see review in White et al. 2000). Disturbances such as fires, insect outbreaks, and floods affect human life and property outside protected areas. Human activities can also affect the magnitude and rate of disturbances entering protected areas. Therefore, managers must be concerned with land use on the lands surrounding protected areas (Hansen and DeFries 2007).

Landscape Ecology

Landscape ecology has emerged since the 1980s as the study of the "causes, consequences, and dynamics of spatial pattern in ecosystems" (Don Falk, personal communication, 2008). Among the elements of pattern are patches (discrete areas of habitat or successional stages), matrix (the contrasting lands surrounding the patches), corridors (linear features that create connections between patches), stepping stones (linear but discontinuous features that can facilitate movement between patches), and barriers (features that form areas of resistance to movement between patches). These elements can be static features of the landscape, such as the position of a river valley, or dynamic, such as the shifting mosaic of burned and unburned patches in a fire-prone landscape. Landscape ecology addresses both the effect of the pattern on process (e.g., the effect of landscape pattern of fuels on fire behavior) and the effect of process on pattern (e.g., the effect of fire in creating a spatial mosaic of different successional states across a landscape). The definition of landscape elements—indeed, even the size of the landscape—will vary with the questions and processes of interest in a particular ecosystem.

In island biogeography, the effect of the matrix (the ocean) is simply a function of the distance between islands and mainland, but the matrix itself is neutral. Landscape ecology allows the matrix to be positive (e.g., resting places, food sources) or negative (e.g., increased predation) for species movements and can both increase and decrease the movement of disturbances. Landscape models also allow variation in patch quality, address the degree of contrast between patches and matrix, and provide a framework for determining edge effects. Spatially explicit landscape models can be run on real maps to provide a framework for analysis of habitat patches, corridors, stepping stones, and barriers to dispersal. Thus, landscape ecology has given us tools to understand the flows of species, energy, nutrients, and disturbances across complex landscapes. Of particular interest here is a focus on matrix lands and the corridors and connections that increase the functional size and decrease the functional isolation of protected areas (Hansen and DeFries 2007). We return to these subjects later.

Large-Scale Planning and Coordination

Three interrelated types of large-scale planning and coordination are needed to meet protected area goals: coordination between separate protected areas to produce both diversity and redundancy in management strategies; planning of the spatial configuration of protected areas into networks, nodes, and corridors; and integrated management across the landscape, managing protected areas and the matrix in which they are situated. We review these three types of large-scale conservation planning in this section. We draw on such initiatives as the Wildlands Project, the Biosphere Reserve Program, the Canadian Conservation First effort, The Nature Conservancy's Conservation by Design, ecosystem management, and the "greater ecosystem" concept (Grumbine 1994; Christensen et al. 1996), and lessons learned from large-scale watershed restoration projects (reviewed by Doyle and Drew 2008).

Coordination between Protected Areas

As discussed in earlier chapters, the future is uncertain, as is predicting the effectiveness of different strategies for responding to climate change and other stressors. The risks associated with such uncertainty may be best managed through diversity and redundancy of management strategies. Diversity is also demanded by the varied goals of parks and wilderness and by the different management emphases that stem from these goals: autonomous nature, ecological integrity, historical fidelity, and resilience (explored in depth in Part II). The complement to diversity of management strategies is redundancy. If a particular approach fails in one place, because of a poor match to environmental conditions or just chance, that approach may be successful elsewhere. Diversity of management strategies and replication of management approaches across the landscape also can increase the opportunity to adaptively learn from management results.

Diversity and redundancy of management strategies must be planned at multiple scales. They can be applied within individual protected areas through different strategies in different parts of a single park or wilderness. More uncommon, more challenging, and even more important is planning for diversity and redundancy at a larger scale, among protected areas. Currently, the diversity and redundancy that exist are the serendipitous result of numerous individual decisions made by different managers, many of whom vary in the importance they attach to the different meanings of naturalness. This ad hoc diversity reflects the subjective interpretation of different management goals, politics, and available resources rather than a large-scale, planned, and deliberate effort that considers the appropriateness of interventions, scale, boundary effects, and ways in which any particular area fits in the larger system of protected areas and the regional landscape.

What is needed is to enlarge the scale at which decisions about where and how to intervene in ecosystem processes are made, to move beyond case-by-case and area-by-area decision making. McCool and Cole (2001) argue that by focusing on the "art of the possible," such area-by-area decision making is often blind to the long term and spatially expansive consequences of decisions. They argue that "a common outcome of incremental decision-making is that even though each manager may be acting responsibly for the area under his or her jurisdiction, the cumulative effect of individual, apparently unrelated decisions may be a situation that was never explicitly intended and is difficult to reverse" (p. 86), the diminution of important values, homogenization of conditions, and suboptimization of the aggregate value and benefit of protected area systems. White and Jentsch (2005) make a similar point about ethics and decisions in the context of patch dynamics. They argue that issues of sustainability are most critical not at the commonly observed scale of individual patches but at the multipatch scale and that "it is at this scale that the essential right or wrong exists: resilience, retention of elements, and continued ability to respond to disturbance are a property of the collection of patches rather than the individual patch" (p. 9).

The challenge thus becomes one of ensuring that individual place-specific decisions are made in the context of larger-scale regional strategies. Managers of protected areas within the same bioregion need mechanisms that will allow the development of regional plans for conserving biodiversity and responding to threats, such as invasive species and climate change. These plans would distribute goals and strategies among individual protected areas in such a manner that when implemented locally they collectively provide diverse and redundant management strategies at the regional scale. Different public agencies would have to collaborate across jurisdictional boundaries and extend the process to include private lands.

An example of this approach is provided by planning efforts by the national forests in North Carolina to preserve hemlocks from an invasive insect, the hemlock wooly adelgid (USDA Forest Service 2005). Mortality of hemlock trees caused by the adelgid (*Adelges tsugae*), both in a number of wilderness areas and on surrounding Forest Service lands, suggested that Carolina hemlock (*Tsuga caroliniana*) and possibly even the eastern hemlock (*Tsuga canadensis*) might become extinct. But the intervention option of injecting insecticide and introducing nonnative beetles, though shown to be successful at allowing infested trees to recover (Cheah and McClure 2002), had its own risks and certainly compromised the wilderness ideal of autonomous nature and untrammeled ecosystems. Ultimately, planners decided on an objective that represented a compromise between intervention and letting nature take its own course. They decided to attempt to maintain reproducing populations of eastern and Carolina hemlock throughout their historic and elevational range. Notably, decisions about where to intervene were determined more by the spatial distribution of stands, in concert with concepts of the metapopulation and minimum viable population size, than by whether the stand was inside or outside a wilderness area.

Networks of Nodes, Corridors, and Buffers

Noss (1983) argues that conservation planning historically took place at the site level and emphasized content (the habitat diversity within a protected area). He called for more emphasis on context (conservation units as parts of larger-scale landscapes and regions) to support seasonal movements and migration patterns of wildlife and the ability of large-scale processes such as fire and flood to occur unimpeded. Noss proposes the idea of ecological networks, with core conservation areas (nodes) connected with corridors (Noss et al. 1999). The function of the core areas is to protect the needs of species that are sensitive to human activities. Corridors may allow reciprocal immigration and genetic exchange, decrease extinction rates, and minimize the effects of catastrophes on populations (Simberloff et al. 1992). The node and corridor approach often incorporates buffer areas surrounding the nodes, an idea that was championed as early as the 1960s in the design of biosphere reserves. Management of the buffer zone should be undertaken with a goal of not threatening preservation of the node. Cores, corridors, and buffers aim to overcome fragmentation, a major threat to biodiversity, increase effective park size, and thereby perpetuate the ecological processes that operate at large spatial and temporal scales (Noss and Harris 1986).

Networks of protected areas (nodes and corridors) have not been validated in practice because most are still in development (Bennett and Wit 2001), and their contribution to conservation success will occur over long time periods. However, such networks have received international attention. Article 8 of the Convention on Biological Diversity calls for the establishment of an international system of protected areas, and globally there are more than 500 biosphere reserves in more than 100 countries. The biosphere reserves contain legally protected core areas, a buffer that supports conservation in the core area, and areas that allow resource use and extraction. The intent is to support sustainability for both biodiversity and local human populations.

Network and corridor plans are also being developed at subcontinental scales. For example, the Wildlands Project has developed large-scale conservation plans (Noss 2003) for four "megalinkages" in North America: the Pacific Megalinkage (Baja California to Alaska), the Spine of the Continent Megalinkage (Mesoamerica to Alaska), the Atlantic Megalinkage (Florida to New Brunswick), and the Boreal Megalinkage (Maritime Provinces to Alaska). Other examples include the Pan-European Ecological Network, which includes core areas, corridors, buffers, and restoration areas spanning fifty-two countries in Europe and northern Asia, and Gondwana Link in southwestern Australia. Most of these plans have yet to be implemented in policy or practice.

Despite substantial research indicating the conservation benefits of networks and corridors in some situations (Schmiegelow 2007), some researchers suggest taking a cautionary approach (Boitani et al. 2007). For example, corridors are species-specific and may be unused or irrelevant to the conservation of some species. Originally conceived as a requirement for migratory animals, corridors are now viewed as critical to the ability of many plants and animals to migrate in response to climate change. Some networks, such as Natura 2000 sites, use existing protected areas as core areas, although there is little evidence that these areas actually represent ideal core areas for biodiversity (Boitani et al. 2007). Connections may allow the spread of diseases and invasive species and undermine the risk spreading that occurs if separate areas are not subject to the same environmental catastrophes. These potential problems suggest that it is critical to articulate specific goals for corridors, then prioritize goals and design corridors for key species and processes.

Integrating Protected Areas and the Matrix

Hansen and DeFries (2007) provide an excellent framework for management of lands surrounding protected areas in ways that will support protected area goals and take into consideration human uses of those lands. Their framework focuses on the following four mechanisms: changes in ecosystem size (a mechanism derived from the concepts of the minimum dynamic area, the species–area relationship, and the maintenance of large predators and hence trophic structure), altered flows of materials and disturbances in and out of the reserve, effects on habitats that support migrations and metapopulation dynamics (e.g., source–sink dynamics), and exposure to human effects across the protected area boundary, such as invasive species, disease, and wild harvest of organisms. Hansen and DeFries (2007) link each of these mechanisms to the following management strategies that would maximize protected area success.

Increasing the effective ecosystem size of protected areas. Barring their actual increase in size through continued land acquisition, effective size can be increased by establishing buffers or zones around protected areas that are managed to support processes and population viability within the protected area.

Altered flows of materials and disturbances in and out of the reserve. Hansen and DeFries (2007) recommend mapping ecological process zones around protected areas to identify the adjacent areas that contribute to these altered flows. For example, fire can move both into and out of a protected area, and the flow of fire in both directions is greatly influenced by restoration activities, fuel levels, and management use of fire on both sides of the protected area boundary. By mapping the ecological process zones, managers obtain a spatial template for cooperative management.

Effects on habitats that support migrations and metapopulation dynamics. The management response to the reduction in critical habitat and migration corridors depends on the species of concern, but the general principle is to determine the important habitat patches outside protected area boundaries and to identify the routes of migration for species dependent on corridors, continuous riparian and aquatic habitats, and stepping stone dispersal.

Exposure to human effects across the protected area boundary, such as invasive species, disease, and wild harvest of organisms. The management response to human effects across the boundaries of protected areas would include regulation of the purposeful and accidental transport of invasive species around protected areas, regional monitoring of disease populations, and methods to ensure sustainable harvest of organisms outside the reserve and prevention of illegal harvest within the reserve.

Facilitating Large-Scale Conservation through Institutional and Policy Change

As the science of conservation at larger scales has matured and protected area managers have recognized the importance of working across jurisdictional boundaries, mechanisms for information exchange have been established. Various working groups meet periodically to exchange ideas and keep each other apprised of plans. For example, the Crown of the Continent Manager's Partnership meets to discuss each unit's plans as they relate to issues of broader concern, such as fire and endangered species. Unfortunately, these efforts to share information, though an important first step, do not always result in changes in management across jurisdictional boundaries.

Key institutional, cultural, and political barriers must be addressed to facilitate cooperation and coordination of ecosystem and landscape-level conservation, to move these multijurisdictional collaborations beyond information exchange. In a study of Ontario Parks, managers claimed they did not have the jurisdiction to pursue the most desirable climate change adaptation strategies (Lemieux et al. 2008). Public land managers are often reluctant to infringe on each other's autonomy and authority, limiting the level of specificity and the kind of accountability involved in such efforts. Furthermore, according to Karkkainen (2002: 23),

> Government agencies often have relatively narrow missions. We are accustomed to managing environmental and natural resource problems one-at-a-time and in isolation from each other, as if pollution control and water supply and fisheries management and habitat conservation

had nothing to do with each other. . . . The lesson of ecosystems as complex dynamic systems is that we need to manage each in the context of the others, and all in the context of the whole, each in its own unique locally/regionally situated ecosystem context. . . . Because different ecological processes operate at different scales, the management "unit" can be difficult to define. Existing institutions—states, counties, federal government—are oftentimes either too big or too small. Moreover, and crucially, beyond *intergovernmental* coordination, the new ecosystem management institutions typically involve *interagency* coordination across functional mission and program lines. Thus we see wildlife management and fisheries officials working side-by-side with land managers, water supply managers, representatives from environmental protection agencies, and forestry and agriculture agency officials.

Specific agency directives to work across boundaries could empower managers to move toward meaningful coordination. Incentive systems, such as additional funding for particular units or promotions for key leaders, could be instituted. Public forums where agencies are held accountable for working at the ecosystem, landscape, or continental scale could increase public pressure.

Ultimately, statutory guidance may be needed to address conflicting policy mandates and mandate coordination across jurisdictional boundaries. Conflicting policy mandates send federal agencies in very different directions. In the Greater Yellowstone Ecosystem, the Bureau of Land Management and the Forest Service have argued that National Park Service policy mandates should not influence management of their lands, despite obvious spillover effects (Lockhart 1991). Nie (2003) argues that clearer legislative priorities are necessary for effective ecosystem-scale management in the Yellowstone area. Furthermore, with the exception of the National Wildlife Refuge System Improvement Act (1997), none of the statutes governing public land management actually mandate cooperation across boundaries.

Finally, policy change can help facilitate private landowner involvement in landscape-scale conservation. Large-scale planning requires meaningful public involvement and engagement with diverse stakeholders (see Chapter 13). Private landowners in particular must be included in conservation efforts in order to create and maintain connectivity at scales larger than protected areas. Conservation easements provide a reasonably effective mechanism to limit development on private lands. However, easements tend to focus on preserving open space by limiting subdivision development and often provide few stipulations regarding habitat for native species. To provide connectivity as conditions change, easements will need to incorporate new provisions that ensure that habitat is provided over the long term.

Conservation in the Greater Yellowstone Ecosystem: A Case Study

Although Yellowstone National Park is one of the largest parks in North America south of Alaska, it provides one of the best examples of attempts to plan and manage at even larger spatial scales. The 890,000-hectare park's boundaries are defined primarily by the occurrence of geothermal features. If boundaries were to be defined by ecological features, such as the movements and dynamics of species such as bison (*Bison bison*) and grizzly bears (*Ursus arctos horribilis*) or the area's fire regime, a much larger area would need to be considered. In this case study we highlight two large-scale efforts: the development of a science-based conservation design and efforts to work across jurisdictional and disciplinary boundaries to further bioregional conservation.

To determine effective park size for Yellowstone National Park, Noss et al. (2002) used a quantitative analysis to develop a conservation plan for 10.8 million hectares, an area more than ten times the size of the national park itself. This area included 2.9 million hectares of protected lands and 7.9 million hectares of matrix, split between public (64 percent) and private ownership (36 percent). Goals were "(1) to represent ecosystems across their natural range of variation, (2) to maintain viable populations of native species, (3) to sustain ecological and evolutionary processes, and (4) to build a conservation network that is resilient to environmental change" (p. 896). They overlaid information on imperiled species and communities; vegetation, geoclimatic, and aquatic environmental variation; and critical habitat and viability analysis for focal species. Using an algorithm that included analysis of vulnerability and irreplaceability, areas were proposed for conservation that would most contribute to the four overall regional conservation goals. Their algorithm also took into consideration the compactness of selected areas to maximize connectivity. One innovation of this project was the inclusion of viability analysis for focal species. The following five species were selected to represent the ecological processes and large area needs of key species: grizzly bear, gray wolf (*Canis lupus*), wolverine (*Gulo gulo*), lynx (*Lynx canadensis*), and elk (*Cervus canadensis*). The inclusion of these species required the recognition of an expanded study area of 31.6 million hectares. This expanded area is about thirty times larger than Yellowstone National Park.

While scientists have been investigating the scales needed to sustain biodiversity, federal agencies within the Greater Yellowstone Ecosystem have been engaged in an experiment in cross-boundary conservation since 1985, through the Greater Yellowstone Coordinating Committee (GYCC). The GYCC consists of eleven lead decision makers who collectively manage the region's two national parks, six national forests, and three national wildlife refuges (14 million of the area's 19 million acres) (Clark 2008) (Figure 12.1). The GYCC exists to coordinate decisions, strategies, and practices across federal jurisdictions on a wide range of issues from elk to

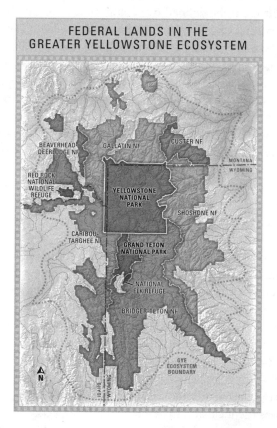

FIGURE 12.1. Large-scale conservation requires meaningful coordination and negotiation between federal and state agencies, private landowners, and local communities. (Map by Bruce Capdeville)

water quality, and from noxious weeds to land use (Clark 2008). In 1991, Keiter and Boyce (394–395) suggested that the GYCC represented a "serious commitment" to the development of institutional mechanisms to coordinate across jurisdictions but that, at that time, meaningful ecosystem conservation was "more a myth than a reality." In the 1990s, the GYCC increasingly embraced the principles of ecosystem management and moved, albeit slowly, toward integration and coordination across management units (Clark 2008).

But according to long-time GYCC scholar Clark, meaningful conservation across boundaries in the Greater Yellowstone Ecosystem has largely gone unrealized. The GYCC tends to focus on noncontroversial issues and delays decision making in favor of continued information gathering. Participating agencies have conflicting legislative mandates; common goals and priorities, where articulated, are usually broad and nonbinding. Because agencies are reluctant to violate each other's autonomy, in some cases they continue to pursue radically different objectives on nearby parcels. A dramatic example is development of oil and gas reserves on Forest Service land alongside biodiversity conservation on Park Service lands (Clark 2008). The GYCC lacks state and local representatives and has little ability to address conservation on private lands, despite the fact that private lands in the area are rapidly transitioning from ranchland to residential development. Perhaps because the GYCC has not been entirely inclusive of local communities, public land management in the area is characterized as "conflict-ridden," and there is a notable local lack of trust for public agencies (Clark 2008).

Still, Clark (2008) argues that the GYCC has tremendous potential to provide meaningful leadership for conservation across boundaries. However, for the GYCC to provide real leadership across jurisdictions it needs clearer goals, mechanisms for making binding decisions, and structural changes that allow collaboration with state and local representatives. The slow progress of the GYCC illustrates the formidable barriers to cross-jurisdictional conservation, barriers that can be diminished only through policy change (addressing conflicting mandates), cultural change (nurturing leaders willing to embrace a larger vision and make it binding), and institutional change (increased engagement with local communities).

Moving toward Conservation at Larger Scales

Conservation of biodiversity and other values typically associated with protected areas requires management at scales larger than individual protected areas. Three kinds of large-scale planning are important: coordination between separate protected areas to produce both diversity and redundancy in management strategies; planning of the spatial configuration of protected areas into networks, nodes, and corridors; and integrated management of protected areas and the matrix in which they occur. Conservation planning at these scales should be process-based and focused on increasing the effective size and connectivity of protected areas, managing disturbance regimes, and allowing for seasonal migrations and adjustment to environmental change. Large-scale conservation projects will inevitably span both human-dominated and nature-dominated areas and involve many institutions. Adaptive management and adjustment to changing circumstances require that science be integrated with decision making (Noss 2003). Large-scale conservation efforts require coordination, openness, transparency, and mechanisms for conflict resolution (Doyle and Drew 2008). Large-scale conservation planning is challenging institutionally, politically, and ecologically, but it is key to the conservation of biological diversity and the capacity for adaptational change.

BOX 12.1. LARGE-SCALE PLANNING AND PROTECTED AREA STEWARDSHIP

- Protected areas are rarely established by ecological boundaries and often are an incomplete sample of the habitats, species, and processes that structure their ecosystems.
- Large-scale conservation planning is needed to restore functions, ensure representation of ecological variation, restore connectivity, increase effective protected area size, and produce adaptable and resilient landscapes.
- Large-scale conservation planning must address coordinating between protected areas, planning networks of protected areas within bioregional landscapes, and managing surrounding lands to increase the effective size and reduce the isolation of the protected areas.
- Large-scale conservation will require changes in polices and coordination between agencies and institutions.

REFERENCES

Bennett, G., and P. Wit. 2001. *The development and application of ecological networks: A review of proposals, plans, and programs.* World Conservation Union (IUCN), Gland, Switzerland.

Boitani, L., A. Falcucci, L. Maiorano, and C. Rodinini. 2007. Ecological networks as conceptual frameworks or operational tools in conservation. *Conservation Biology* 21:1414–1422.

Cheah, C. A. S., and M. S. McClure. 2002. *Pseudoscymnus tsugae* in Connecticut forests: The first five years. Pp. 665–691 in B. Onken, R. Readeon, and J. Lashomb, eds. *Proceedings, Hemlock Woolly Adelgid in the Eastern United States symposium*. Rutgers University, East Brunswick, NJ.

Christensen, N. L., A. M. Bartuska, J. H. Brown, S. Carpenter, C. D'Antonio, R. Francis, J. F. Franklin, et al. 1996. The report of the Ecological Society of America Committee on the Scientific Basis for Ecosystem Management. *Ecological Applications* 6:665–691.

Clark, S. G. 2008. *Ensuring Greater Yellowstone's future: Choices for leaders and citizens*. Yale University Press, New Haven, CT.

Diamond, J. M. 1975. The island dilemma: Lessons of modern geographical studies for the design of nature preserves. *Biological Conservation* 7:129–146.

Doyle, M., and C. Drew. 2008. *Large-scale ecosystem restoration: Five case studies from the United States*. Island Press, Washington, DC.

Foppen, R. P. B., J. P. Chardon, and W. Liefveld. 2000. Understanding the role of sink patches in source–sink metapopulations: Reed warbler in an agricultural landscape. *Conservation Biology* 14:1881–1892.

Grumbine, R. E. 1994. What is ecosystem management? *Conservation Biology* 8:27–38.

Hansen, A. J., and R. DeFries. 2007. Ecological mechanisms linking protected areas to surrounding lands. *Ecological Applications* 17:974–988.

Karkkainen, B. C. 2002. *Collaborative ecosystem governance: Scale, complexity, and dynamism*. Paper presented at the conference Saving Nature: Theories, Tools, and Strategies in Environmental Conservation, October 2000 at the University of Virginia School of Law, Charlottesville.

Keiter, R B., and M. S. Boyce. 1991. *The Greater Yellowstone Ecosystem*. Yale University Press, New Haven, CT.

Lemieux, C. J., D. J. Scott, R. G. Davis, and P. A. Gray. 2008. *Climate change, challenging choices: Ontario parks and climate change adaptation*. University of Waterloo, Department of Geography, Waterloo, ON.

Levins, R. 1969. Some demographic and genetic consequences of environmental heterogeneity for biological control. *Bulletin of the Entomological Society of America* 15:237–240.

Lockhart, W. J. 1991. "Faithful execution" of the laws governing Greater Yellowstone: Whose laws? Whose priorities? Pp. 49–64 in R. B. Keiter and M. S. Boyce, eds. *The Greater Yellowstone Ecosystem: Redefining America's wilderness heritage*. Yale University Press, New Haven, CT.

MacArthur, R. H., and E. O. Wilson. 1967. *The theory of island biogeography*. Princeton University Press, Princeton, NJ.

McCool, S. F., and D. N. Cole. 2001. Thinking and acting regionally: Toward better decisions about appropriate conditions, standards, and restrictions on recreation use. *The George Wright Forum* 18(3):85–98.

Newmark, W. 1987. A land-bridge island perspective on mammalian extinction in western North American national parks. *Nature* 325:430–432.

Nie, M. 2003. *Beyond wolves: The politics of wolf recovery and management*. University of Minnesota Press, Minneapolis.

Noss, R. F. 1983. A regional landscape approach to maintain diversity. *BioScience* 33:700–706.

Noss, R. F. 2003. A checklist for wildlands conservation network designs. *Conservation Biology* 17:1270–1275.

Noss, R. F., C. Carroll, K. Vance-Borland, and G. Wuerthner. 2002. A multi-criteria assessment of the irreplaceability and vulnerability of sites in the Great Yellowstone Ecosystem. *Conservation Biology* 16:895–908.

Noss, R. F., and L. D. Harris. 1986. Nodes, networks, and MUM's: Preserving diversity at all scales. *Environmental Management* 10:299–309.

Noss, R. F., J. R. Strittholt, K. Vance-Borland, C. Carroll, and P. Frost. 1999. A conservation plan for the Klamath–Siskiyou ecoregion. *Natural Areas Journal* 19:392–411.

Pickett, S. T. A., and J. N. Thompson. 1978. Patch dynamics and the design of nature preserves. *Biological Conservation* 13:27–37.

Schmiegelow, F. K. A. 2007. Corridors, connectivity and biological conservation. Pp. 251–262 in D. B. Lindenmayer and R. J. Hobbs, eds. *Managing and designing landscapes for conservation: Moving from perspectives to principles*. Blackwell, Malden, MA.

Schoenwald-Cox, C. 1983. Guidelines to management: A beginning attempt. Pp. 414–446 in C. Schoenwald-Cox, S. M. Chambers, B. MacBryde, and W. L. Thomas, eds. *Genetics and conservation*. Benjamin-Cummings, New York.

Shelford, R. E. 1933. Ecological Society of America: A nature sanctuary plan unanimously adopted by the society, December 28, 1932. *Ecology* 14:240–245.

Shugart, H. H. 1984. *A theory of forest dynamics*. Blackburn, Caldwell, NJ.

Simberloff, D., J. A. Farr, J. Cox, and D. W. Mehlman. 1992. Movement corridors: Conservation bargains or poor investments? *Conservation Biology* 6:493–504.

Taylor, M., and P. Figgis. 2007. Protected areas: Buffering nature against climate change—Overview and recommendations. Pp. 1–12 in M. Taylor and P. Figgis, eds. *Protected areas: Buffering nature against climate change*. Proceedings of a WWF and IUCN World Commission on Protected Areas symposium, June 18–19, 2007, Canberra. World Wildlife Fund–Australia, Sydney.

Terborgh, J., J. A. Estes, P. Paquet, K. Ralls, D. Boyd-Heger, B. J. Miller, and R. Noss. 1999. The role of top carnivores in regulating terrestrial ecosystems. Pp. 39–64 in M. Soulé and J. Terborgh, eds. *Continental conservation*. Island Press, Washington, DC.

Turner, G. M., W. H. Romme, R. H. Gardner, R. V. O'Neill, and T. Kratz. 1993. A revised concept of landscape equilibrium: Disturbance and stability on scaled landscapes. *Landscape Ecology* 8:213–227.

USDA Forest Service. 2005. *Environmental assessment: Suppression of hemlock woolly adelgid infestations*. National Forests in North Carolina, Asheville.

White, P. S., J. Harrod, W. Romme, and J. Betancourt. 1999. The role of disturbance and temporal dynamics. Pp. 281–312 in R. C. Szaro, N. C. Johnson, W. T. Sexton, and A. J. Malk, eds. *Ecological stewardship*, Vol. 2. Elsevier Science, Oxford.

White, P. S., J. Harrod, J. L. Walker, and A. Jentsch. 2000. Disturbance, scale, and boundary in wilderness management. Pp. 27–42 in S. F. McCool, D. N. Cole, W. T. Borrie, and J. O'Loughlin, eds. *Wilderness science in a time of change conference*. Vol. 2: *Wilderness within the context of larger systems*. Proceedings RMRS-P-15-VOL-2. Rocky Mountain Research Station, Ogden, UT.

White, P. S., and A. Jentsch. 2001. The search for generality in studies of disturbance and ecosystem dynamics. *Progress in Botany* 62:399–450.

White, P. S., and A. Jentsch. 2005. Developing multipatch environmental ethics: The paradigm of flux and the challenge of a patch dynamic world. *Silva Carelica* 49:93–106.

Chapter 13

Planning in the Context of Uncertainty: Flexibility for Adapting to Change

F. STUART CHAPIN III, ERIKA S. ZAVALETA,
LEIGH A. WELLING, PAUL DEPREY, AND LAURIE YUNG

The only thing constant in life is change.

—*François de La Rochefoucauld*

Throughout human history, people have interacted with and shaped the ecosystems of which they are a part (Turner et al. 1990; Diamond 2005). In the last 50 years, however, human activities have changed Earth's climate and ecosystems more rapidly and extensively than in any comparable period of the last 10,000 years (Steffen et al. 2004; Millennium Ecosystem Assessment 2005; Intergovernmental Panel on Climate Change [IPCC] 2007; Chapter 4, this volume). Many of the fundamental controls over social-ecological systems, such as climate, economic globalization, and land cover change, show persistent or accelerating trends over time. Consequently, the structure and dynamics of ecosystems will continue to change, probably even more rapidly than in the recent past. As described earlier in this book, these changes challenge traditional goals for protected areas, such as preserving naturalness, and the approaches that managers typically use to meet those goals.

To remain effective, protected area planning must change to meet the needs of this new context. The directional nature of many of these changes poses a serious challenge for the stewardship of protected areas because there is no historical reference point that serves as an appropriate target. Furthermore, uncertainty about future conditions and the outcomes of management interventions requires more flexible and nimble planning practices. Deciding whether and how to intervene in the face of uncertainty is particularly challenging, and managers need well-defined tools to help guide such decisions.

In this chapter, we first examine current typical planning approaches and make the case that current approaches are not sufficient for planning in the face of change. We then describe five general approaches to planning that can help managers understand uncertainty and be more adaptive: identification and monitoring of protected area changes in the context of regional and global change, sensitivity analysis to assess the social-ecological consequences of observed trends, scenarios to explore consequences of plausible future changes and intervention options, fostering of adaptive capacity to generate the flexibility to respond to and shape future change, and adaptive management that engages stakeholders to test understanding and implement potential solutions. Scenario planning is explored in detail through a case study in Joshua Tree National Park.

The Limitations of Traditional Planning Approaches

Planning is "a process of identifying a desired future and determining the pathway (or set of pathways) to it" (McCool et al. 2007: 3). Park and wilderness planning is usually based on a rational comprehensive model that assumes agreement on objectives, scientific certainty, and the availability of data to support a decision (Lachapelle et al. 2003). Such planning assumes that there is a single "best" answer that can be determined through careful analysis (Karkkainen 2002). In reality, however, protected area management is characterized by competing goals and claims, lack of scientific consensus, and uncertainty about future conditions and the outcomes of management actions.

Most planning methods are based on assumptions that climate, species distributions, and ecological processes are stable (Kareiva et al. 2008), and most plans identify targets or reference conditions that are based on the historical range of variability (Dixon 2003). For example, fire planning for low-elevation ponderosa pine forests in the western United States often uses historic fire regimes as a target for the frequency and intensity of fire. Similarly, wolf (*Canis lupus*) reintroduction in the Lamar Valley of Yellowstone National Park is judged successful to the extent that ecosystems return to past vegetation and soil conditions. Such planning assumes that past states are knowable and that future conditions will be similar to past conditions. But in the context of large but uncertain future changes, desired future conditions that are defined in terms of a specific past may be impossible to attain (see Chapter 4). For example, preventing species range shifts in the context of climate change may limit opportunities for adaptation.

Both temporal and spatial scales are problematic in most plans. Once finalized, plans are expected to guide the next 10–20 years. Managing agencies sometimes adhere strictly to such plans, even as conditions change and the plan becomes untenable, or recognize that the plan is outdated and proceed in the absence of any deliberate planning process, without adequate consideration of the uncertainty associated with persistent directional trends or novel landscape interactions. Because planning horizons often cannot adapt to the pace or extent of change, the mismatch between the length or area of the plan and the timing and spatial range of actual changes render plans less effective and relevant.

Traditional planning also privileges technical experts and efficient, predictable decision processes. Public involvement is legally mandated, but it can be cursory and procedural and is not always fully embraced by decision makers. Agencies sometimes subscribe to a "decide–announce–defend" model whereby they propose a specific action, then vigorously defend it throughout the planning process. Such an approach seriously limits opportunities for dialogue and collaboration, frustrating interested members of the public. Political conflict over protected area management is common, and stakeholders may find few avenues for meaningful involvement in decision making.

For all these reasons, existing protected area planning tools are insufficient for addressing the challenges associated with uncertainty and rapid change. Planning tools for an unknown but rapidly changing future must detect and address changes occurring at scales ranging from the individual protected area to their more human-dominated surroundings and beyond, because many of the inevitable surprises result from interactions across these scales. Goals and the specificity of targets must be considered carefully. Planning must be ongoing, with multiple time horizons, short- and long-term goals, frequent updates, and flexibility to adjust to unexpected changes. Planning processes must engage stakeholders in all phases and aspects of decision making.

Understanding Uncertainty

Crafting an effective planning process requires recognition of multiple types of uncertainty. Some changes are quite likely to occur, although the rates and patterns of change are uncertain. For example, given the long lifetime of CO_2 in the atmosphere, past human CO_2 emissions commit the planet to a trajectory of decades to centuries of continued climate warming and subsequent impacts on Earth's systems (IPCC 2007). Consequently, species will probably continue to migrate poleward and upward in elevation in response to this warming, and wildfire in some regions may continue to increase (Kasischke and Turetsky 2006).

Managers often seek to avoid large threshold changes in structure and function, because such changes may be difficult or impossible to reverse. Consequently, consideration of potential threshold changes is often missing from the planning process because of high uncertainty in whether or when they will occur. For example, loss of glaciers from Glacier National Park or declining permafrost in arctic and boreal ecosystems will cause permanent hydrologic changes, but the timing is quite uncertain. Identifying plausible thresholds and their likely triggers and consequences are key components of planning in the face of uncertainty.

Finally, there are "unknown unknowns." These surprises will inevitably occur, even though their exact nature cannot be predicted. Loss of the American chestnut (*Castanea dentata*) from introduced chestnut blight, for example, eliminated one of the forest dominants in the deciduous forest of the eastern United States. Although, by definition, it is impossible to plan specific responses to unanticipated events, more nimble and responsive planning processes provide flexibility to respond to whatever uncertain future might unfold. The greater the uncertainty of future conditions, the greater the value of planning that fosters resilience, flexibility to shift in mid-course, and adaptability.

Identifying and Monitoring Change

Monitoring to document rates and patterns of change is essential for planning. Nonetheless, time and funding constraints often limit monitoring of changes in protected areas or of trends in drivers of change in the surrounding region. An effective monitoring strategy must characterize threats and changes and identify key response variables that would be sensitive indicators of changes that are likely to occur or would have profound ecological or social impacts if they did occur. Managers are particularly reluctant to commit to long-term monitoring of slowly changing control variables, such as biodiversity, that are often the best indicators of fundamental ecosystem change.

A common impediment to monitoring is the all-or-nothing perception that monitoring must be complex and precise to be useful (Holloran 2006). Monitoring should be simple, flexible, consistent, and spatially extensive, but it need not always be precise, complex, or comprehensive (in variables measured, design, or statistical robustness). It can often be qualitative and opportunistic. For example, if staff who are tasked with trail and waste management are encouraged to report ecological threats and trends, such as arrival of invasive species or evidence of forest health problems, this both engages staff in the larger issues of the protected area and establishes a consistent basis for spatially dispersed monitoring.

Tools for Planning in the Context of Uncertainty

In this section we outline several tools for planning in the context of uncertainty and potentially rapid but unpredictable change. These tools work in concert and include sensitivity analysis, scenario planning, adaptive capacity, and adaptive management.

Sensitivity Analysis

In ecological models, sensitivity analysis is used to determine how sensitive the outcome of a model is to particular assumptions about each piece of it. For instance, how sensitive is a prediction of sustainable fish populations to assumptions about year-to-year stability in adult survival? In planning, sensitivity analysis can be used with scenario-based planning to test and validate assumptions and interactions between factors that make up a scenario storyline or to evaluate the robustness of potential policy directions or management options to unanticipated futures. If option A is chosen because it has favorable outcomes and low risks under current conditions, how much can the system change in various directions before option A no longer works? We illustrate this later in a case study of Joshua Tree National Park.

Sensitivity analysis can be done at various levels of sophistication depending on the skills and resources available. Also, sensitivity analyses can be conducted using either quantitative or qualitative methods. A valuable first cut at sensitivity analysis is to survey local experts, including scientists, local naturalists, farmers, and ranchers, about observations of past sensitivities to extreme events or of anticipated sensitivities to expected changes. Where consensus emerges, this provides a valuable framework for the design of monitoring programs or the development of models. Where differences of opinion emerge, this can help identify research needs.

Scenario Planning

Scenario-based planning provides a starting point for exploring policy options that work across a range of possible futures or intervention options to assess the risks and opportunities. Scenarios represent plausible futures based on our understanding of past and current trends. They are not predictions because of the uncertainties that surround the future. Rather, they can be used to explore alternative possibilities for the future, as illustrated in the case study presented later.

Scenarios are most effective when used comparatively to explore the logical consequences of differences in assumptions about how the world works or the social-ecological consequences of alternative policy options. The choice of scenarios to be explored is one of the most important steps in designing a scenario-based approach to planning. Because it is generally not feasible to explore more than four or five scenarios at a time (Ogilvy and Schwartz 2004; Carpenter et al. 2006), planners must carefully choose the scenarios on which to focus. Some select a scenario based on its likelihood of occurrence, the risk of catastrophic consequences, or its utility in identifying management tools to address projected changes. Other approaches involve identifying the drivers of change that are most uncertain and carry the largest consequences and exploring the intersection of these drivers. Scenarios that explore changes that are extremely unlikely may be considered if the consequences are very large. Consequences that cannot be altered by managers, such as those driven by external forces such as climate change, may be important to consider for adaptation planning. For example, coastal zone managers need to consider what to do in the face of certain sea level rise. Scenarios that have negligible effects are of little use in planning.

The greatest shortcomings of scenarios include the tendency to over-emphasize the present and our ability to predict the future, lack of knowledge about the assumptions that underlie models of the system and the consequences of missing information, the challenge of capturing the complexities of social-ecological interactions and feedbacks, and failure to consider "wildcards," the unknowable surprises that are an increasingly common feature of social-ecological systems (Peterson et al. 2003; Carpenter et al. 2006). Clearly, scenarios should be treated as plausible futures rather than predictions.

Scenarios can also be useful in addressing more focused questions at finer geographic scales using qualitative or quantitative modeling. In Alaska, for example, scenarios have been developed to guide managers in decisions about how to address expected increases in extent and frequency of wildfire associated with climate warming. Wildlife refuge managers know that wildfire leads to improved moose (*Alces alces*) habitat 15–30 years after fire but that the winter range of caribou (*Rangifer tarandus*) generally takes 80–100 years to recover because of the slow colonization and growth of lichens, the major winter food of caribou (Maier et al. 2005; Chapin et al. 2008). Historically, refuge managers have allowed wildfires to burn, except adjacent to communities, because wildfire is a natural process essential to the long-term sustainability of the boreal forest. However, models of the effects of projected warming on wildfire and caribou habitat suggested that the expected future increase in wildfire extent might eliminate most caribou winter range within the current century, endangering one of the key species that the refuge was established to protect (Rupp et al. 2006). Scenarios that changed levels and configuration of fire suppression activities identified a new management strategy that substantially reduced the risk of catastrophic reduction in caribou winter range.

Adaptive Capacity

Simply knowing that changes are occurring or that the risks of likely changes are large is usually insufficient to precipitate action. Adaptive capacity is the capacity of a social-ecological system to learn, cope, innovate, and adapt to changes that might occur. Said another way, it is the capacity of the system to be resilient to both expected and unexpected future changes (specific and general resilience, respectively; see Chapter 9 for more detail). The adaptive capacity of protected areas depends on ecological diversity at genetic, species, and landscape levels. This diversity provides essential building blocks that allow the system to sustain its current functioning under a wide range of future conditions. Management to sustain biodiversity requires actions in both protected areas and surrounding areas (see Chapter 12). For example, if some species or functional types disappear with climate change, porous linkages with adjacent landscapes, both natural and human-dominated, allow recolonization and maintenance of functions that might otherwise be lost. Finding the right balance between fostering mobility of desirable species and minimizing invasion of aggressive exotic species requires an understanding of the ecology of potential colonizing species and factors that control their migration (Chapin et al. 2007; Millar et al. 2007).

Adaptive capacity also depends strongly on the flexibility of protected area management to cope with, adapt to, and shape change. This is particularly challenging where agency culture has fostered rigid protocols and rules in an effort to meet a restricted set of management objectives such as timber harvest or species conservation. In a world without variability and surprise, an inflexible management approach provides short-term predictability and efficiency. However, these potential advantages are outweighed by a low capacity to respond to change, as uncertainty and variability increase. In the Pacific Northwest, for example, decades of sustained-yield timber management led to an agency culture that proved inflexible in dealing with changing societal goals (e.g., protection of endangered species), climate, and wildfire (Trosper 2003; Swanson and Chapin 2009).

Adaptive capacity can be enhanced by rewarding flexibility and innovation within agencies and by broadening communication and interaction between agencies, nongovernment organizations, and citizen groups. Informal networks of citizens and interest groups often provide effective informal venues for policy discussions that might not be possible in more formal settings structured by historical objectives, approaches, and power relationships. Overlap in management responsibilities between institutions can also provide flexibility to adapt to change. If an agency and a conservation nongovernment organization both monitor population changes of rare or invasive species, for example, this effort is less likely to be eliminated as a result of sudden shifts in funding or priorities in either group.

Adaptive Management

In a rapidly changing world, it is not feasible to postpone actions until there is certainty about the trajectories of change or the consequences of alternative management options. Instead, managers must learn by doing without destroying future options. This process of adaptive management recognizes that uncertainty is inevitable and builds learning into the management process. Under adaptive management, decisions are provisional, and there is flexibility to adapt to change. Such management requires constant feedback from monitoring that documents the outcomes of management actions (Karkkainen 2002). Active adaptive management involves deliberate management experiments to test competing hypotheses (Kareiva et al. 2008). Adaptive management is dynamic and flexible because learning and adjustment are built into planning. For example, the transport (assisted migration) of immobile species to higher elevations in regions where rapid climate change threatens current populations is a management experiment that might reduce the vulnerability of threatened species, but it might also have undesirable outcomes because of unanticipated species interactions. Assessing the desirability of management experiments requires careful assessment of the risks and opportunities associated with potential outcomes. For example, policies that allow some small insect outbreaks to occur may be preferable to preventing insect damage by pesticide application, which increases the continuity of susceptible stands and increases the likelihood of a catastrophic outbreak (Holling and Meffe 1996).

For adaptive management to be effective, the planning process itself must be adaptive, revised in the light of new learning and experimentation. Planning would thus need to experiment with different time horizons and spatial scales, mechanisms for involving stakeholders, and ways of conducting management experiments. The inevitable uncertainty in future conditions usually makes it impossible to know which of several policy options will lead to preferred outcomes, so experimentation with a single management option is generally insufficient. Coordination of adaptive management experiments in protected areas with planning efforts to address similar issues through management interventions in the surrounding landscape provides opportunities to enhance learning. In the Pacific Northwest, for example, the Forest Service manages 85 percent of their lands with conservation and resource protection as primary goals. On the remaining 15 percent of their lands timber is harvested, using several silvicultural approaches. Finally, timber companies harvest trees on short rotations on their private lands. Carefully designed comparisons provide opportunities to learn about the relative performance of each approach, each of which has risks and uncertainties. Planning in the context of this entire mix of land uses, rather than focusing on a single protected area strategy, broadens the dialogue about synergies and trade-offs and brings into focus issues related to interactions between protected lands and the surrounding landscape.

Adaptive management is not a simple process, and many complex practical issues emerge that are specific to each situation, particularly in addressing uncertainty. Legal and policy constraints may restrict the planning approaches that can be used, at least in the protected areas. Finally, policies that are designed to be flexible to adapt to unexpected future outcomes may lack the teeth for effective protection under current conditions. Uncertainty challenges decision makers to develop policies that can adapt to changing conditions without becoming vulnerable to abusive, wide-ranging interpretations.

Engaging Stakeholders

To build political capital and socially acceptable plans, adaptive management must involve stakeholders in meaningful ways. Engagement with stakeholders must be dynamic and long term, because decisions will need to be provisional. Because adaptive management requires flexibility and a certain amount of managerial discretion, public support is critical. Planning processes must engage stakeholders as co–problem solvers at all stages, from initial framing and goal setting to evaluation of monitoring results and adjustment of management actions. Engagement of diverse stakeholders increases the likelihood that the full breadth of synergies and trade-offs will be considered and that there will be broad support for policies and actions that emerge. This process can draw on diverse sources of local, traditional, and formal knowledge systems that widen the consideration of goals and approaches available to address complex problems. Increasing the breadth of stakeholders also increases the likelihood that interactions between protected areas and surrounding, more human-dominated landscapes will be considered. Broad participation also generally taps or initiates informal networks and bridging organizations that communicate issues and capabilities to other organizations and agencies with related concerns. This increases the breadth of resources and effort that can be entrained to address complex problems. Meaningful public involvement increases transparency and builds the trust necessary for flexibility and experimentation (Lachapelle et al. 2003).

Scenario Planning in Joshua Tree National Park: A Case Study

Because of the potentially dramatic impacts of climate change on the iconic species that Joshua Tree National Park protects, the National Park Service convened a group of fifteen scientists, managers, and educators in 2007 to explore the use of scenario planning as a structure for decision making in an uncertain future. Their efforts provide an excellent example of the use of scenario planning to assist protected area managers in planning in the context of uncertainty and potentially rapid change. The Joshua Tree effort generally followed five steps outlined by Peterson et al. (2003) but also drew from the work of Ogilvy and Schwartz (2004), the Climate Impacts Group (2007), and Liu et al. (2008). General steps and outcomes are given in this case study.

Defining the Focal Question

The group settled on a fairly general question to focus the scenario planning effort: How can national park managers respond to climate change impacts? To answer the question, the group needed to understand the legal and policy context of the park and know which resources and ecological processes might be affected and how. Joshua Tree National Park was created by the California Desert Protection Act in 1994. The park was established to preserve natural resource values in this unique natural landscape, perpetuate in their natural state significant and diverse ecosystems of the California desert, protect and preserve historical and cultural values of the California desert, provide for public recreation while protecting and interpreting the natural and cultural resources, and enhance scientific research in undisturbed ecosystems. Two quite distinct desert ecosystems are represented in the park. The Mojave Desert occurs at higher elevations to the northwest, and the Colorado, or Sonoran, Desert occurs at lower elevations to the southeast.

Identifying External Drivers

In addition to climate change, the group identified two social drivers that have the greatest impact on the ability of the park to meet the goals set forth in its enabling legislation: park budgets and conservation values. Park budgets are made up largely of operational funds (this is the largest and most stable part of the budget) and project funds (these funds are more changeable over time and focus on meeting short-term exigencies such as repaving roads or conducting ecosystem restoration). Conservation values are articulated in enabling legislation but can also change through time. For example, Joshua Tree was originally designated as a national monument (a cultural resource preservation effort) but was renamed a national park with passage of the 1994 California Desert Protection Act, which established wilderness and identified additional (natural resource) conservation values to be associated with the park. These changes in national and regional conservation values have modified the park and will undoubtedly continue to do so.

Exploring Internal Dynamics

To define plausible alternative pathways into the future, the group used tables to document what is known about the internal relationships and dynamics of the system (including means, trends, extremes, seasonality, and associated uncertainties across a range of climate drivers and ecosystem responses). Access to climate change data, research and monitoring results, predictive models, and the associated scientific uncertainties were critical inputs at this stage. The group also used conceptual diagrams to map the relationships and feedbacks between various components to help identify where the park is most vulnerable to change.

Building and Testing Scenarios

The goal in creating scenarios is not to identify the most likely scenario but to flesh out qualitatively different alternative plots or storylines for how the future might unfold such that the range of variability is adequately represented. Testing scenarios involves looking for internal consistency and plausibility for how the story is expected to play out and validating the robustness of assumptions. Three scenarios were developed for Joshua Tree, based largely on the uncertainty in amount and timing of precipitation under a warming future. It should be pointed out that it was not possible to fully test the assumptions for the scenarios given the limited amount of time available at the workshop. They are briefly presented here to illustrate the process.

In the "Summer Soaker" scenario, annual precipitation does not change, but less rain falls in winter and spring and more falls during summer monsoons. This scenario was constructed to be consistent with IPCC scenario B1, which assumes rapid reduction in greenhouse gas emissions. Because summer rains favor annual native grasses, this could help to curtail

FIGURE 13.1. The expansive and dense Cholla Garden is located in the transition zone between the lower Colorado (Sonoran) Desert and the higher Mojave Desert. Climate change could alter the specific conditions that support its abundant growth. (Photo by the National Park Service)

invasion by nonnative vegetation. Warmer temperatures drive vegetation communities to move upslope, causing the Mojave ecosystem to be reduced and the Sonoran ecosystem to expand. As the transition zone between the two ecosystems is altered, features that occur along the transition would be affected. For example, a popular feature in this zone is Cholla Garden, a dense growth of cholla cactus (*Opuntia* sp.) (Figure 13.1). Warmer temperatures and erosion from intense summer rains may threaten the distinctive nature of the Garden. In addition, as the Mojave ecosystem is reduced, some of the species native to this system, such as the bighorn sheep (*Ovis canadensis*) or the relic namesake species, the Joshua tree (*Yucca brevifolia*), would probably become isolated or could be lost altogether from the park. Other species, such as the desert tortoise (*Gopherus agassizii*), may do better as their vegetative browse (summer native grasses) increases, although increased summer moisture may exacerbate the upper respiratory tract disorder in tortoises.

"When It Rains It Pours" is a scenario in which extreme precipitation events are common, especially in winter, and often follow summers of extreme drought. This scenario was constructed to be consistent with IPCC scenario A1B, an intermediate emission scenario. Chief concerns are flash floods and erosion, causing debris dams in canyons that blow out, increased disruption of traffic and other visitor activities, safety concerns, and higher costs for infrastructure maintenance and emergency response. Flooding and erosion would destroy many easily damaged archaeological sites, although new sites may be uncovered. Conditions would also enhance a positive feedback loop involving drought, invasion by exotic grasses, and fire, converting the system to a grassland ecosystem with a more extreme fire regime (i.e., summer drought kills off native annuals, and heavy winter rains follow that promote growth of exotic invasive grasses that act as fuels for fire, which fertilizes the ground for nonnative annuals). This shift stresses many native species that occupy small niches and do not thrive on less nutritious nonnative grasses.

In the "Dune" scenario, the park experiences increasing temperature with persistent dryness and drought. This scenario was constructed to be consistent with IPCC scenario A1F1, which assumes extremely rapid global warming. Wind increases in frequency and intensity. Change in vegetation habitats caused by drought and high temperatures leads to a significant loss of woody species. Fire spread increases because of increased nonnative vegetation fuel load and increased winds. Eventually, as more fires occur and consume available vegetation, fire occurrence declines. This was considered the most devastating of the three scenarios because it would cause very substantial loss of vegetative cover due to fire, water mining, wind, and higher temperatures, and wind-induced dune formations in the Pinto Basin.

Policy Screening

Ultimately, this step brings the most value to park managers because common elements across the scenarios are identified and the costs and benefits of actions can be evaluated. As with the previous step, the following list is not comprehensive; the list illustrates some of the common issues that were identified across the three scenarios and some suggestions for actions that could help managers cope with them.

Loss of Mojave Desert species and expansion of Sonoran. Regardless of the rate, warming will drive plant and animal communities to adapt or shift their location. For high-priority species such as the Joshua tree, park managers may consider relocating species to higher elevations to promote their survival. Other actions might include conducting research on hybrid species that could survive warming and strengthening seed banks and nurseries to support future restoration.

Increased size and severity of wildland fire. The incidence of fire increases in all scenarios,˙so the park needs to consider now how best to manage for this. Exploring different options for fuel breaks on the landscape was one idea discussed.

Invasion of nonnative plants. Combating exotic grasses and other nonnative plants (existing and potential new invasions associated with climate change) will be important for reducing the threat of fire and maintaining native species. Funding must be obtained to support staff efforts in responding immediately to these threats.

Putting the Plans into Action

The scenario planning exercise at Joshua Tree revealed some common issues that are relevant across a range of possible climate change conditions. The park must now determine how best to incorporate this information into park plans and operations. In 2009, the park will embark on formal development of a general management plan (GMP), the outcome of which will guide long-range (15–20 years) decisions about appropriate uses, desired futures, and infrastructure and facility development. The plan provides an opportunity to develop management actions related to the scenarios explored.

Also, although scenario planning for Joshua Tree National Park has not involved members of the public to date, the GMP process provides an opportunity to facilitate a broader dialogue about possible futures and management options. Scenario planning is a particularly effective tool for engaging stakeholders in joint discovery and mutual learning, for incorporating nonexpert knowledge into plans, and for building public support and political capital. Ideally, such planning exercises should engage the public at all stages, from scenario building to evaluation of management actions.

The results from this exercise may also be incorporated into National Park Service climate change adaptation strategies at the national and regional scales. The National Park Service is in the process of outlining adaptation options and goals and expects that many individual parks will engage in scenario planning to develop park-specific adaptation plans that address protection of key resources under future climate change conditions.

Planning in the Context of Change

In the past, protected area planning was predicated on the assumptions that conditions were stable, that experts could select targets with some certainty, and that public involvement, though required, was not critical to good decision making. But given potentially dramatic yet somewhat unpredictable ecological changes, planning must evolve to account for uncertainty. In the face of change and uncertainty, protected area planning must become more dynamic, flexible, and responsive, attentive to change and adaptive in its response to change.

Protected areas are best managed in ways that accommodate appropriate changes rather than simply seeking to prevent inappropriate changes. Efforts to prevent all changes are likely to create a system that is far from equilibrium with its social and environmental controls. This may increase the risk of large catastrophic changes. Planning in the context of change therefore requires the identification and monitoring of potential changes,

---

BOX 13.1. EFFECTIVE PLANNING IN THE FACE OF CHANGE AND UNCERTAINTY

- Effective planning entails understanding different types of uncertainty, including the following:

    Changes that are likely to occur, although rates and patterns are uncertain

    Threshold changes and their triggers and consequences

    Changes that are plausible, although likelihood and triggers are uncertain

    Unknown unknowns or surprises
- To plan in the context of uncertainty, managers can

    Monitor change to document what changes have occurred

    Use sensitivity analysis to test and validate specific assumptions and interactions

    Engage in scenario planning to explore a range of futures or intervention options and the associated risks and opportunities

    Foster adaptive capacity, with a focus on flexibility and innovation

    Practice adaptive management by learning through management experiments and monitoring of outcomes
- Stakeholders should be engaged in planning as active participants in defining and solving problems, to draw on diverse sources of knowledge, foster linkages across the landscape, develop networks, increase trust and transparency, and build political capital and public support.

an understanding of their potential social and ecological consequences, and an assessment of the trade-offs associated with these changes. Because many of these trade-offs entail value judgments that may differ between stakeholders, meaningful engagement of stakeholders in the planning process is important for managing both temporal change and spatial interactions.

REFERENCES

Carpenter, S. R., E. M. Bennett, and G. D. Peterson. 2006. Scenarios for ecosystem services: An overview. *Ecology and Society* 11(1):29. Retrieved from www.ecologyandsociety.org/vol11/iss1/art29/.

Chapin, F. S. III, K. Danell, T. Elmqvist, C. Folke, and N. L. Fresco. 2007. Managing climate change impacts to enhance the resilience and sustainability of Fennoscandian forests. *Ambio* 36:528–533.

Chapin, F. S. III, S. F. Trainor, O. Huntington, A. L. Lovecraft, E. Zavaleta, D. C. Natcher, A. D. McGuire, et al. 2008. Increasing wildfire in the boreal forest: Causes, consequences, and pathways to potential solutions of a wicked problem. *BioScience* 58:531–540.

The Climate Impacts Group. 2007. *Preparing for climate change: A guidebook for local, regional, and state government.* King County and ICLEI: Local Governments for Sustainability, Seattle.

Diamond, J. 2005. *Collapse: How societies choose or fail to succeed.* Viking, New York.

Dixon, G. E. 2003. *Essential FVS: A user's guide to the forest vegetation simulator.* USDA Forest Service, Forest Management Service Center, Fort Collins, CO.

Holling, C. S., and G. K. Meffe. 1996. Command and control and the pathology of natural resource management. *Conservation Biology* 10:328–337.

Holloran, P. 2006. Measuring performance of invasive plant management efforts. Pp. 12–13 in *Research and management: Bridging the gap.* Proceedings of the California Invasive Plant Council Symposium 10. October 5–7, 2006, Rohnert Park, CA.

Intergovernmental Panel on Climate Change. 2007. *Climate change 2007: The physical science basis.* Cambridge University Press, Cambridge.

Kareiva, P., C. Enquist, A. Johnson, S. H. Julius, B. Petersen, L. Pitelka, R. Shaw, and J. M. West. 2008. Synthesis and conclusions. Pp. 9-1–9-66 in S. H. Julius and J. M. West, eds. *Preliminary review of adaptation options for climate-sensitive ecosystems and resources: A report by the U.S. Climate Change Science Program and the Subcommittee on Global Change Research.* U.S. Environmental Protection Agency, Washington, DC.

Karkkainen, B. 2002. Collaborative ecosystem governance: Scale, complexity, and dynamism. *Virginia Environmental Law Journal* 21:189–243.

Kasischke, E. S., and M. R. Turetsky. 2006. Recent changes in the fire regime across the North American boreal region: Spatial and temporal patterns of burning across Canada and Alaska. *Geophysical Research Letters* 33:doi:10.1029/2006GL025677.

Lachapelle, P. R., S. F. McCool, and M. E. Patterson. 2003. Barriers to effective natural resource planning in a "messy" world. *Society and Natural Resources* 16:473–490.

Liu, Y., M. Mahmoud, H. Hartmann, S. Stewart, T. Wagener, D. Semmens, R. Stewart, et al. 2008. Formal scenario development for environmental impact assessment studies. Pp. 145–162 in A. J. Jakeman, A. A. Voinov, A. E. Rizzoli, and S. H. Chen, eds. *Developments in integrated environmental assessment*, Vol. 3. Elsevier, Amsterdam.

Maier, J. A. K., J. Ver Hoef, A. D. McGuire, R. T. Bowyer, L. Saperstein, and H. A. Maier. 2005. Distribution and density of moose in relation to landscape characteristics: Effects of scale. *Canadian Journal of Forest Research* 35:2233–2243.

McCool, S. F., R. N. Clark, and G. H. Stankey. 2007. *An assessment of frameworks useful for public land recreation planning.* General technical report PNW-GTR-705. USDA Forest Service, Pacific Northwest Research Station, Portland, OR.

Millar, C. I., N. L. Stephenson, and S. L. Stephens. 2007. Climate change and forests of the future: Managing in the face of uncertainty. *Ecological Applications* 17:2145–2151.

Millennium Ecosystem Assessment. 2005. *Ecosystems and human well-being: Synthesis.* Island Press, Washington, DC.

Ogilvy, J., and P. Schwartz. 2004. *Plotting your scenarios.* Global Business Network, Emeryville, CA.

Peterson, G. D., G. S. Cumming, and S. R. Carpenter. 2003. Scenario planning: A tool for conservation in an uncertain world. *Conservation Biology* 17:358–366.

Rupp, T. S., M. Olson, J. Henkelman, L. Adams, B. Dale, K. Joly, W. Collins, and A. M. Starfield. 2006. Simulating the influence of a changing fire regime on caribou winter foraging habitat. *Ecological Applications* 16:1730–1743.

Steffen, W. L., A. Sanderson, P. D. Tyson, J. Jäger, P. A. Matson, B. Moore III, F. Oldfield, et al. 2004. *Global change and the Earth system: A planet under pressure.* Springer-Verlag, New York.

Swanson, F. J., and F. S. Chapin III. 2009. Forest systems: Living with long-term change. Pp. 149–170 in F. S. Chapin III, G. P. Kofinas, and C. Folke, eds. *Principles of ecosystem stewardship: Resilience-based natural resource management in a changing world.* Springer, New York.

Trosper, R. L. 2003. Policy transformations in the US forest sector, 1970–2000: Implications for sustainable use and resilience. Pp. 328–351 in F. Berkes, J. Colding, and C. Folke, eds. *Navigating social-ecological systems: Building resilience for complexity and change.* Cambridge University Press, Cambridge.

Turner, B. L. II, W. C. Clark, R. W. Kates, J. F. Richards, J. T. Mathews, and W. B. Meyer, eds. 1990. *The earth as transformed by human action: Global and regional changes in the biosphere over the past 300 years.* Cambridge University Press. Cambridge.

Chapter 14

Wild Design: Principles to Guide Interventions in Protected Areas

ERIC S. HIGGS AND RICHARD J. HOBBS

> Design is not just what it looks like and feels like. Design is how it works.
>
> *—Steve Jobs*

Managers of protected areas intervene in ecosystem processes regularly. They regulate trail use, decommission access roads, revegetate trampled alpine meadows, remove invasive species, and alter fire regimes. Thus, one of the primary questions concerning the appropriateness of intervention has been answered countless times through routine and exceptional actions. The debate is not about whether intervention is appropriate in some places and in some situations but how managers should intervene intelligently and when they should back away. The challenges for managers are becoming more complex with the stressors of invasive species and rapid environmental changes. The historical ground underneath our management philosophies, policies, and actions is shifting. Managers are being asked to make timely judgments about interventions that stray far from experience and from the comfort zone of traditional ethical practice. For instance, should species near the limits of their present range be assisted in moving to new locations? How much intervention is justified, and on what basis? What is the risk of doing harm in attempting to correct a problem?

Much of this book explores the challenge of interventions in protected area ecosystems: when they should be taken, for what purpose, and with what desired outcome. In this chapter, we apply a version of design practice, wild design, to the question of intervention and outline a framework for managers based on seven principles (clarity, fidelity, resilience, restraint, respect, responsibility, and engagement) that focus on understanding of ecological characteristics and ethical challenges. Design is the intention and planning behind any action. Wild design refers to intentions and plans that recognize and support free-flowing ecological processes. Thus, there is a critical tension between unrestrained processes (wild) and human intervention (design) in wild design. We believe this tension is implicit in many of the challenges faced by contemporary protected area managers and that a comprehensive framework is needed to successfully adapt to changing conditions. Our proposal is not for a stepwise procedure for protected area managers but instead a framework in which to fit a wide variety of contemporary and emerging management challenges.

Protected areas are an exquisite test for wild design. There are many motivations for designating protected areas, but in all cases the ecological and cultural qualities of protected areas demand particular respect. For example, deciding how to design a restoration project in the black oak woodlands of Yosemite National Park should be at least as reverent an undertaking as the restoration of a great historic building. Managers are willing to go more slowly, attend more carefully to the qualities of the thing or ecosystem, invest greater energy in finding durable approaches, and recognize all the while that posterity is at stake. Therefore, protected areas provide the greatest challenge for working out how to intervene appropriately, if in fact intervention is ultimately appropriate. They are also places where there are the greatest restrictions on and resistance to resolute intervention. Finally, there are places too remote, too fragile, or the subject of long-term study that should simply be left alone (see Chapter 6). The framework of wild design is appropriate in many places but not everywhere.

The Evolution of Design

Design is anathema to most protected area managers. It implies a heavy hand of intervention, as though it were possible or even reasonable for people to design wild places. But each decision to intervene in protected area ecosystems is a decision of design and should be acknowledged as such. The notion of design fits with traditional views about the process of realizing built structures, products, advertising, and logistics. However, the evolution of design from built structures to complex systems provides an opportunity for ecologists and protected area managers to undertake a deliberate, systematic, and ethically grounded approach to intervention.

Design theory and practice arose in the twentieth century as a response to emerging industrial and technological developments. The need for better and more integrated systems and products promoted the new fields of industrial and symbolic design. Buchanan (1992) argues that design arose as a series of evolutionary stages beginning with symbolic and visual communication (advertising, marketing) and progressing through artifacts (industrial, fashion), activities, and services (logistics) to complex systems for living and working.

The design of ecosystems and landscapes has a long history in landscape architecture and allied practices and runs roughly parallel to the larger pattern of development in design practice. Early notions of landscape design applied to human-dominated environments (e.g., Forman and Godron 1986; Dorney 1989; McHarg 1995) are now being translated to efforts to manage and restore landscapes for conservation and other purposes (Lindenmayer and Hobbs 2007). In many ways, landscape design was a response to the preoccupation with aesthetic aims in landscape architecture. Beginning in the 1960s, McHarg and others suggested that ecological processes, patterns, and structures be included in designs (Buchanan 1992; Borgmann 1995; Orr 2001) and that ecological principles be the center around which design revolves (Johnson and Hill 2002; Nassauer and Opdam 2008). Especially with increasing awareness of the consequences of rapid ecological and environmental change, there is the need for deliberate actions by protected area managers to reinforce existing values (wilderness, ecological integrity, historical fidelity) or find salutary forms of adaptation (resilience). What is needed, therefore, is a distinctive approach to design that embeds ecological and ethical principles.

The Distinctiveness of Wild Design

Wild design is a formulation of design principles and practices intended explicitly for managers who are compelled for various reasons (legal requirements, loss of critical ecosystem components) to intervene in ecological systems. Wild design originated as one of four keystone concepts in Higgs's (2003) formulation of ecological restoration and was offered as a reconciliation of two divergent qualities of restoration: an emphasis on free-flowing ecological processes (wildness; see Chapter 6 for more detail) and deliberate intervention to meet human objectives, even if these objectives are putatively based on an understanding of ecosystems. Restoration is only one type of intervention, but it is also representative of a broad array of practical and ethical issues involved with intervention.

Wild design operates on the insight that a tension between intervention and wildness requires maintenance. Too much emphasis on human values will deflect attention away from ecological integrity, and pure consideration of wildness will miss the critical participation of people in intervention. Following Nassauer and Opdam (2008), we propose that design is the interface between ecological intervention and ecological learning (Figure 14.1). There are two continuous cycles that flow through wild design. There is a broader suite of ecological knowledge that encompasses both scientific inquiry and what arises from traditional knowledge, literature, art, and other forms of understanding. Wild design is one of the steps in the cycle that includes theory, experiments, and assessments. The other cycle, ecological intervention, includes goal setting, wild design, implementation, and monitoring. Thus, wild design serves as the process in common between knowledge acquisition (science and traditional knowledge) and ecological intervention. In other words, it is the activity that translates knowledge into practice.

Interventions in protected area ecosystems are based substantially on ecological goals. There are many reasons why managers intervene, including reducing harm to people and infrastructure, recovering ecological integrity through prescribed fire, translocating critically imperiled species, restoring complete historical assemblages, eliminating or controlling invasive species, and enhancing adaptation to a changing climate. The scale of

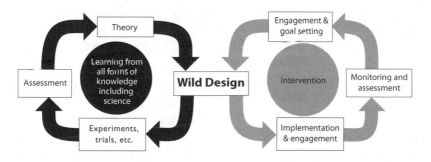

FIGURE 14.1. How wild design connects knowledge acquisition and ecological intervention. (Modified from Nassauer and Opdam 2008)

intervention ranges from the simple act of placing a fence around a trampled patch of vegetation or rerouting a trail to restoring the historic fire return interval over a vast landscape.

There are almost as many policies and procedures as there are interventions, and each jurisdiction operates within a distinct socioeconomic and political reality. Explicit recognition of the role of people points to the following insights. First, a basic ecological and technical understanding of management problems is important, but it is only one step along the path to successful intervention. Second, recognizing that intervention takes place in a social and political arena means that human values are important in arriving at a reasonable outcome. The appreciation and negotiation of differing perspectives are a key role for contemporary protected area managers. Third, participation by the public in all dimensions of a project is increasingly important to achieve support. Thus, the policy environment suggests that wild design needs to incorporate principles and guidelines that give rise to public engagement, ethical procedures, and a fine balance between ecological and practical considerations.

The Process of Wild Design

The function of wild design, therefore, is to provide an overarching set of principles and guidelines for practice, not necessarily to supplant local and regional considerations. New national guidelines in Canada, for example, offer a comprehensive approach to recovering ecological integrity in national parks and provide an overall schematic for intervention (Parks Canada 2008). The process begins with a general consideration of values, in this case the ones set by Parks Canada (efficient, effective, and engaging), and more generally the six International Union for Conservation of Nature protected area management categories. This sets the boundary conditions for the project, including the appropriate balance between human and ecological considerations. Next, the problem is defined through field and documentary research. Goals for the project are established based on the principles, and specific objectives are honed to address the carefully defined problem. Guidelines in the Parks Canada document are tagged to specific types of intervention. From there, the steps are entirely familiar: Develop and implement a plan and ensure adequate monitoring and adaptive management. Of course, the devil is always in the details, but this approach is generally consistent with the expectations of most protected area managers.

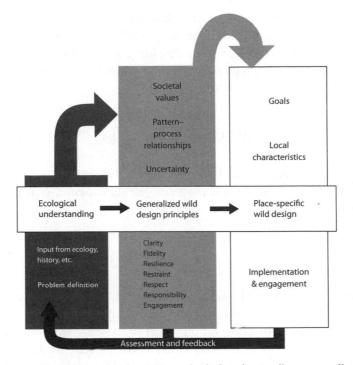

FIGURE 14.2. The relationship between ecological understanding, generalized wild design principles, and place-specific wild design. (Modified from ideas presented in Nassauer and Opdam 2008)

However, wild design calls for a more intensive consideration of cultural and ecological values, which we embed in a series of design principles. We also ensure that there is sufficient feedback from practice to principles to theory to encourage appropriate adaptations for complex projects in fast-moving systems. Again, we turn to Nassauer and Opdam for inspiration (Figure 14.2). The process begins with ecological understanding. Data derived from ecological studies, including ecological and human history, are gathered to provide a detailed account of the ecosystem's characteristics and threats. The second stage moves from ecosystem knowledge to the generalized conditions for wild design by using a series of principles that incorporate societal values and uncertainty. This middle stage is crucial and distinctive. It compels the manager to use design principles as a filter between knowledge and action. The third stage, place-specific wild design, involves goal setting, engagement with stakeholders, attentiveness to local characteristics, and project implementation. There is assessment and feedback throughout the process to allow for adaptation. Many of the latter steps in the Parks Canada approach described earlier would be place-specific wild design. Thus, we are providing a more general framework onto which many local, regional, state or provincial, and national intervention guidelines can be attached.

Wild Design Principles

Wild design is underpinned by seven principles. Ecosystems vary, of course, and so do specific intervention practices, but these design principles provide orientation throughout.

Clarity

All interventions should be based on clear goals and objectives to ensure the transparency of the values guiding those interventions. Clarity illuminates the ambitions of managers on behalf of the ecosystem and provides measurable targets against which to assess the project. Such targets can be expressed as a range of potential outcomes or as a series of prospective trajectories. In Gulf Islands National Park, for example, managers might establish as a goal the control of feral goat populations on Saturna Island (Golumbia 2006). The eradication of the goat population is highly controversial among residents of the island, but so are the impacts of the goat population. Objectives might include additional research on the goat population and on the values of island residents and visitors, demonstration exclosure plots to show the effects of removing trampling and herbivory, and public engagement in the research and management process (e.g., an advisory committee). Clear goals and objectives expose underlying value assumptions and allow the development of appropriate and more socially acceptable interventions.

Fidelity

Wild design should be faithful, to some degree, to the ecosystem. Fidelity entails careful historical research to understand the past conditions of the system, including major disturbances, and to assess these past conditions against present functions, structures, and patterns. Although it is seldom appropriate to return an ecosystem to specific historical conditions, a careful review of the past reveals subtle characteristics of an ecosystem that can be critical to successful intervention. It is also vital that historical references be used to elucidate changes in an ecosystem undergoing effects from climate change. A system may be tilting outside previously defined historical ranges of variability, but even this is helpful information. The principle of fidelity is further developed in the Redstreak case study explored later in this chapter.

Resilience

Designs for resilience are necessary to ensure that autogenic functioning is restored to an ecosystem and that an ecosystem has appropriate resources to cope with external perturbations including climate change. Under conditions of rapid change it is important to anticipate as much as possible and adapt to surprises that will make design challenging. Surprise will come in the form of ecosystems that remain stable despite significant change, those that form hybrid systems based on native and new species, and entirely novel or no-analog ecosystems (see Chapter 4 for more detail). The role of wild design is to respond to these changes with flexible interventions and to ensure ecological integrity. Recent developments indicate a move toward considering resilience as a key element in effective restoration. For instance, Jannsen (2006) produced a practical guide for ecosystem restoration in New Zealand that has bush resilience and landscape resilience as primary goals. The purpose of the guide, in Jannsen's words, is to allow practitioners to "score your site's functional elements, which helps you assess its strengths and weaknesses, and develop a management plan to restore resilience to our native bush patch" (p. 18). Similarly, broad-scale conservation initiatives such as Gondwana Link in Australia now have resilience as one of their primary organizational ideas.

Restraint

Restraint is exercised at all times in the design process and implementation of restoration projects. Less intervention is better than more. Knowing when to back away and allow ecological processes to take over is paramount, as is acknowledging where it is inappropriate to intervene, for example in places where the primary goal is wildness (see Chapter 6). Wildfire managers in the Canadian Rocky Mountain national parks, Banff, Jasper, Kootenay, and Yoho, recognize the challenge of putting historical fire conditions back on a vast landscape. Site-specific fire prescriptions would impose too much on budgets and require an intrusive infrastructure to implement. Using natural breaks (e.g., cliffs and ridges) and artificial breaks (e.g., a managed cut swath), it is possible to create capping units, particular areas that in the event of a lightning ignition can be left to burn without intervention. It is tempting to bring even large ecosystems under control, but it is often practical and appropriate to step back and encourage wild processes.

Respect

The manager must be attentive to ecological process and aware that interventions are always proxies for assumptions about what is appropriate to a particular ecosystem. Frank Egler comments that "ecosystems are not only more complex than we think, they are more complex than we can think" (Barbour 1996: 247). The better we understand ecosystems, the more capable we are of respectful action. On Macquarie Island, a World Heritage Site south of Australia, the reduction of one invasive species, cat (*Felis catus*), led to an unexpected increase in herbivory by another invasive species, rabbit (*Oryctolagus cuniculus*). Thus, the elimination of cats, which had predated burrowing seabirds, produced a rapid rise in rabbit populations despite control actions (myxoma virus). These actions were conducted within an integrated pest management framework, but the consequences of rabbit herbivory led to island-wide ecological consequences (Bergstrom 2009). So often, the challenges that appear straightforward are more complicated than we think, and even those anticipated as complex throw up unexpected hurdles.

Responsibility

Wild design brings to the fore the significant responsibility for success and failure that attend ecological intervention. Responsibility includes wide knowledge of techniques and projects, operating according to high ethical standards and striving to allow ecosystems to flourish instead of becoming monuments to human ambition (Allison 2007). Design in this sense provides a check on zealous or ambitious interventions. A critically imperiled mammal, the Vancouver Island marmot (*Marmota vancouverensis*), teetering on the brink of survival in 2001, provides an insightful example. With fewer than seventy-five individuals extant and only twenty-five in the wild, an elaborate recovery program was initiated. The release of captive-reared individuals into historical habitat proved more challenging than anticipated. High rates of predation by golden eagles (*Aquila chrysaetos*), wolves (*Canis lupus*), and cougars (*Puma concolor*) compelled adaptive interventions to temporarily restrain and eliminate local predators (Vancouver Island Marmot Foundation, www.marmots.org). These stopgap measures have worked so far, and the population is climbing slowly. The entire recovery process has been expensive in terms of time and effort, has engaged a wide variety of techniques (captive rearing, wild releases, habitat modification, and predator management), and has courted significant ethical issues (specifically predator control). The intensity and long-term nature of this recovery program demonstrates the sense of responsibility that scientists and managers have for bringing back this species from the brink. Because of the risk of losing imperiled species or ecosystems, interventions should be approached with a sense of ethical duty and profound responsibility.

Engagement

Engagement is the strong reciprocal tie people form with ecosystems through first-hand experience. Wild design involves the engagement of practitioners with the ecosystems in which they intervene and, where applicable, engagement with a wider public in these interventions. More than mere participation in decision making, engagement forms when people connect physically and emotionally with an ecosystem and associated human communities. This is particularly evident with restoration practice and is widely described in the literature (Mills 1995; Gobster and Hull 2000; Light 2002). Engagement, or focal practice (Higgs 2003), connects an individual to a particular place and also enlarges awareness of protection, conservation, and restoration of other places.

The restoration of Beacon Hill Park in downtown Victoria, British Columbia provides an example of effective engagement. A remnant coastal Douglas fir (*Pseudotsuga menziesii*) ecosystem in this urban park was compromised by a century of heavy human use and invasive species in the densely vegetated portions of the site. City of Victoria staff, community volunteers, and a graduate student from the University of Victoria teamed up to recover some of the integrity of this ecosystem by educating the public, removing invasive species, and planting and seeding native species. In 2006, university students enthusiastically volunteered to remove invasives and plant native species. Their direct participation in the project engendered a feeling of connection to the ecosystem and, for many, a commitment to stay involved in the project over the long term. This kind of engagement encourages much-needed involvement in restoration projects (and interventions more generally) and also helps people form a more profound connection to ecosystems.

Troubles with Design

Together, the seven principles form a framework for effective intervention. Application of these principles begins with questions that probe the technical and ethical issues underlying potential interventions. See Table 14.1 for sample questions.

TABLE 14.1. Wild Design Principles and Sample Management Questions

Principle	Guideline	Sample Questions
Clarity	Clear goals and objectives to ensure the transparency of values.	Have goals and objectives been established? Have value claims been made explicit? Are goals and objectives consistent with the interests of those concerned with the intervention?
Fidelity	Entails careful historical research to understand past conditions and to assess these past conditions against present functions, structures, and patterns.	What is known historically about the ecosystem? Have all sources of knowledge been explored? What signals in the contemporary ecosystems can be inferred from historical information?
Resilience	Ensure that autogenic functioning is restored and that the ecosystem has appropriate resources to cope with external perturbations.	What are the functional requirements of the ecosystem? How much continued intervention or management will be needed? Are long-term experiments and simulations being undertaken to elucidate the character of the ecosystem?

TABLE 14.1. Continued

Principle	Guideline	Sample Questions
Restraint	Less intervention is better than more.	Are the means in place to assess the social and ecological impacts of an intervention? Is the line between too little and too much intervention clear (i.e., at what point is less intervention inadequate to meet agreed goals)? Is precaution central to intervention strategies?
Respect	Interventions are always proxies for assumptions about what is appropriate to a particular ecosystem.	Are scientists, managers, and concerned members of the public aware that interventions are simply the best present-day approximation of what is best for the ecosytem and that the assumptions underlying these approximations may shift?
Responsibility	Responsibility includes wide knowledge of techniques and projects, operating according to high ethical standards, and striving to allow ecosystems to flourish.	Are intervention practitioners properly trained? Is there professional conduct in place? Is it clear that the qualities of the ecosystem come before human interests (although there is room for human values too)?
Engagement	Strong reciprocal ties are needed between people and ecosystems to ensure successful and durable interventions.	Have people been brought into the process of designing the intervention early on? Is the role of the concerned public substantial? Who is accountable for the outcome of the intervention? Is community support strong and growing?

This chapter would be missing a central element if we did not include a critical assessment of wild design. After all, *design* is an awkward term to be using in relation to protected areas. The central trouble with design as a metaphor and in practice is that it openly avows and inscribes human beliefs into ecosystems, a process that has caused many of the problems harming ecosystems. There are ways of countering this problem: by viewing design as a process and product, by understanding the open quality of most ecosystems, and by enlarging the possibilities for genuine human engagement.

The first and most evident response to a critique of wild design is to understand design as both process and product. We create designs that are appreciated as artifacts or systems, but we engage in design procedures that lead to specific ends. In the Beacon Hill Park example, the product is a partially restored coastal Douglas fir ecosystem in an intensely urbanized area. Such a product would be unthinkable without extensive public support. This was achieved through seemingly endless meetings and negotiations and a robust volunteer program. The proponents stepped back from the problem and realized that public participation and political support were vital to recovery of the ecosystem. Recognition of these additional elements, particularly ones that slowed down progress on ecological goals, led to a more robust project. Understanding that design is both process and product, and that the former matters as much as and sometimes more than the latter, helps to distinguish wild design from mere meddling.

Second, we embed in the design process an awareness of ecosystems as open, evolving systems that in many cases are more complex than humans can readily understand. Such awareness instills appropriate humility and encourages design practice that is open as well as flexible and adaptable. This humility prevents decision makers from rushing headlong into aggressive assisted migration of threatened Garry oak (*Quercus garryana*) on southern Vancouver Island, for example, because moving species north or south beyond their present range may have unintended consequences. The act of thinking this through carefully, balancing risk and benefits, and proceeding with caution will yield projects that stand the test of time (e.g., a comprehensive network of common garden experiments could produce robust results within the changing biogeographic envelope for various species). Casting such challenges as design encourages a systematic look at open-ended problems. One can argue that this is simply common sense. However, using wild design as a framework maintains a focus on long-term, adaptive thinking.

Third, it is important to overcome the superficiality that can accompany design practice. Such superficiality originates in the professionalization of design and the tendency to create prescribed and formalized rules, procedures, and design grammars for ecosystems that warrant greater creativity. For example, this is a risk with the new national Principles and Guidelines for Ecological Restoration in Canada. Taken too literally, the guidelines could provide a rigid cookie-cutter approach to intervention, which would run counter to the ideas of adaptive management built into the document. The antidote for superficiality is depth. Depth is achieved by ensuring that a primary professional obligation of decision makers is nurturing engagement, which comes about through emphasis on personal and community focal practices (Borgmann 1995; Higgs 2003, 2005) and also by ensuring that general principles for intervention are translated into place-appropriate practices. The principle of engagement is critical in wild design. Engagement takes place when people make bodily and social connections to a place. Opportunities for participation can spark epiphanies, personal realization, and shared collective values. Once such connections are made, there is little risk of superficiality.

The final question is whether wild design is appropriate in all places. No, definitely not, and especially in places where wildness is a primary goal. The problem is in finding the right balance between precaution and action. In the twentieth century, the hands-off approach was used in managing wilderness, while the cumulative human impact on ecosystems grew dramatically. Interventions in protected areas have been ad hoc and haphazard. Modern technological capacity has resulted in an overwhelming confidence in our ability to handle sophisticated challenges, including the restoration of damaged ecosystems. A recent shift to thinking of ecosystems in terms of the services they provide risks pushing the pendulum toward action and away from precaution. Global change necessarily alters perceptions both of what is wild and what should be valued in protected areas. Thus, intervention becomes more attractive and more likely, a mindset that will only strengthen as the consequences of global environmental change become more apparent (Hobbs and Cramer 2008).

The Redstreak Restoration Project: A Case Study

Managers in Kootenay National Park have wrestled with acute ecological and human risk in forested ecosystems. Located on the west side of the Continental Divide of the Canadian Rocky Mountains immediately west of the more famous Banff National Park, Kootenay experiences heavy visitation and intensive use of ecosystems near highways and access roads, including the popular Redstreak campground near the main west entrance of the park and within view of the town of Radium Hot Springs. Nearby forests are typical of those found in the Rocky Mountain Trench, a dry low-elevation valley running north to south between the Purcell and Rocky Mountains. A century of fire suppression, reduced fire incidence, and probably climate change have resulted in forests that are densely packed with coniferous trees and susceptible to extreme fire events. Should a fire ignite in these forests under favorable circumstances, not only would the ecosystems experience the intense heat of crown fire, but human safety would be at risk. Some kind of intervention was needed, but what kind and how much?

Although the origins of the Redstreak Restoration Project predate the formal development of wild design, the planning and implementation activities are an excellent illustration. The location of the Redstreak ecosystem put it very much in the public eye. Experienced managers Rick Kubian and Alan Dibb understood that a bold approach was needed to restore the open Douglas fir forests typical of the Redstreak site. The design included recovery of the open forest structure and a fireguard for the town. Of particular interest was a herd of Rocky Mountain bighorn sheep (*Ovis canadensis*) that had habituated to the grassy verges of roads and the town, an arrangement that led to high mortality and human–wildlife conflict. Restoring the open forest communities would encourage foraging away from built-up areas and would connect with adjacent restoration projects being undertaken just outside the park. The critical first step in the wild design process was met: providing clarity in goals and revealing management and community values.

Kubian and Dibb worked closely with collaborators in the public and private sectors and communicated with hundreds of people formally and informally, putting into practice the principle of engagement, providing clarity about objectives, and exposing management and community values. Working with the Mountain Legacy Project, which produced dozens of repeat photographic pairs based on early twentieth-century survey photographs, and with forest and fire history research by Rob Walker, Robert Gray, and others, managers pieced together an idea of the historical character of the site and ecological processes. They became aware of the mixed-severity fire regime in these ecosystems and the heterogeneous quality of the stand structure. Recognizing the impact of shifting climate, they adopted a flexible approach aimed at ensuring long-term resilience of the forest ecosystems. They were also well aware that their interventions were approximations of what might have happened historically, indicating respect for the ecosystem.

Initial treatment called for selective thinning of the "dog hair" stand of young trees to meet three key objectives: reduce dangerous forest fuel levels around the campground, create a fireguard on the east side of the campground, and restore an open landscape with patches of forests and grasslands. Selective thinning was undertaken in 2003 and followed by a burn in 2005. A low-intensity burn was successfully completed in 2009 to recreate the more frequent historical fire processes.

The results of the restoration were dramatic. In a study of camper reactions to the project, many were taken aback initially by the visual impact of the intervention. Good signage and interpretive programs helped to explain the rationale for the project. Within 2 years, many who visited the campground were unaware of the significant changes and were encountering for the first time an open forest more representative of historical ecological conditions. Original objectives were met, including some change in movement patterns and almost immediate increased use of treated areas by sheep in spring and fall. The response of the community and visitors was largely positive, sufficiently so that more ambitious restoration projects are being planned for larger areas behind the campground. Restraint in this case came about mostly through the use of the Redstreak project as a pilot for potential larger programs in the park. Once outside the immediate vicinity of the town, designs may involve less intervention, where ecological processes can operate largely without significant human involvement. This is consistent with the management of backcountry areas of the park that remain isolated from regular human activity and have suffered fewer direct impacts.

Where is the line drawn between wild design and more conventional intervention? A focus on the many values that condition myriad public and management views of Redstreak allowed a thoughtful and durable approach to intervention. Clear goals, consideration of historical conditions, restraint, and respect helped, too. In the end, Kubian, Dibb, and their colleagues demonstrated a high level of responsibility in making difficult decisions about intervention. The model of wild design is new and requires extensive field testing. So far, it fits well in circumstances where there are likely to be disagreements about the future of ecosystems, which includes just about any ecosystem in an era of rapid environmental change.

Wild Design as Ethical Intervention

Decisions about whether, when, and how to intervene should never be made lightly, especially when the stakes are high. A changing climate and the increasing risk of invasive species and habitat loss will force us to consider interventions more often. Because interventions necessarily involve decisions to design, to intentionally manipulate ecosystem features to achieve specific goals, a framework to guide such decisions is critical. Learning how to intervene respectfully in protected areas, to become responsible wild designers, will increase managers' ability to address rapidly changing conditions effectively. The seven principles described in this chapter are offered as a starting point for developing wild design into a full-fledged practice for protected areas. Unprecedented uncertainty imposes special responsibilities to act resolutely and with humility. This suggests an approach that embeds ethnical norms in the practice of intervention and compels managers to ask questions that continuously reflect back on practice. If there is a single principle that warrants the greatest attention for managers of protected areas, it is engagement. Engagement encourages direct involvement and participation, which has the advantage of creating a constituency or community of support for interventions. The more people understand about an ecosystem, the more they are likely to comprehend the diverse values that typically shape any intervention. Managers will have the greatest long-term success in coping with the unprecedented qualities of ecological and social change in coming decades—novel ecosystems and novel responses—if there is widespread public understanding and support.

BOX 14.1. WILD DESIGN

- Intervention imposes responsibilities best met by a framework that embeds ethical practices.
- All interventions are designs.
- Wild design acknowledges the interplay of human intervention and ecological processes, patterns, and structures.
- Wild design incorporates seven principles for effective practice: clarity, fidelity, resilience, restraint, respect, responsibility, and engagement.
- Although wild design is widely applicable in contemporary protected areas in the face of rapid environmental and ecological change, there remain places (remote, fragile, long-term study) where it is not appropriate.

REFERENCES

Allison, S. K. 2007. You can't not choose: Embracing the role of choice in ecological restoration. *Restoration Ecology* 15:601–605.

Barbour, M. 1996. Ecological fragmentation in the fifties. Pp. 233–255 in W. Cronon, ed. *Uncommon ground: Rethinking the human place in nature.* W.W. Norton, New York.

Bergstrom, D. M. 2009. Indirect effects of invasive species removal devastate World Heritage Island. *Journal of Applied Ecology* 46:73–81.

Borgmann, A. 1995. The depth of design. Pp. 13–22 in R. Buchanan and V. Margolin, eds. *Discovering design: Explorations in design studies*. University of Chicago Press, Chicago.

Buchanan, R. 1992. Wicked problems in design thinking. *Design Issues* 8:5–21.

Dorney, R. S. 1989. *The professional practice of environmental management*. Springer-Verlag, New York.

Forman, R. T. T., and M. Godron. 1986. *Landscape ecology*. Wiley, New York.

Gobster, P., and B. Hull, eds. 2000. *Restoring nature: Perspectives from the social sciences and humanities*. Island Press, Washington, DC.

Golumbia, T. 2006. A history of species introductions in Gwaii Haanas and Gulf Islands national park reserves in British Columbia, Canada: Implications for management. *Transactions of the Western Section of the Wildlife Society* 42:20–34.

Higgs, E. S. 2003. *Nature by design: People, natural process and ecological restoration*. MIT Press, Cambridge, MA.

Higgs, E. S. 2005. The two culture problem: Ecological restoration and the integration of knowledge. *Restoration Ecology* 13:1–6.

Hobbs, R. J., and V. A. Cramer. 2008. Restoration ecology: Interventionist approaches for restoring and maintaining ecosystem function in the face of rapid environmental change. *Annual Review of Environment and Resources* 33:39–61.

Jannsen, H. J. 2006. *Bush vitality assessment: Growing common futures*. Department of Conservation, Wellington, NZ.

Johnson, B., and K. Hill. 2002. *Ecology and design: Frameworks for learning*. Island Press, Washington, DC.

Light, A. 2002. Restoring ecological citizenship. Pp. 153–172 in B. Mintecr and B. P. Taylor, eds. *Democracy and the claims of nature*. Rowman & Littlefield, Lanham, MD.

Lindenmayer, D. B., and R. J. Hobbs, eds. 2007. *Managing and designing landscapes for conservation: Moving from perspectives to principles*. Blackwell Scientific, Oxford.

McHarg, I. L. 1995. *Design with nature*, 2nd ed. Wiley, New York.

Mills, S. 1995. *In service of the wild: Restoring and reinhabiting damaged land*. Beacon, Boston.

Nassauer, J. I., and P. Opdam. 2008. Design in science: Extending the landscape ecology paradigm. *Landscape Ecology* 23:633–644.

Orr, D. 2001. *The nature of design: Ecology, culture, and human intention*. Oxford University Press, Oxford.

Parks Canada. 2008. *Principles and guidelines for ecological restoration in Canada's protected natural areas*. Retrieved October 13, 2009 from www.pc.gc.ca/eng/docs/pc/guide/resteco/index.aspx.

Chapter 15

A Path Forward: Conserving Protected Areas in the Context of Global Environmental Change

LAURIE YUNG, DAVID N. COLE, AND RICHARD J. HOBBS

> In times of change, learners inherit the earth, while the learned find themselves beautifully equipped to deal with a world that no longer exists.
>
> —*Eric Hoffer*

Climate change, habitat fragmentation, atmospheric pollution, and invasive exotic species are changing every ecosystem on Earth. Even protected areas do not provide a safe haven from such changes. And yet the role that protected areas play in conserving native biodiversity and ecological processes is becoming ever more critical. Are park and wilderness managers prepared to respond effectively to the challenges presented by global environmental change? Management intervention is likely to become increasingly pervasive, but is there sufficient policy guidance regarding where and when intervention is needed? Is there adequate guidance regarding what interventions should be taken and what the desired outcomes of those interventions should be?

In this book, we have explored the concept of naturalness, the guiding principle behind protected area conservation in the United States and for many of the countries that have adopted and adapted U.S. models. We have argued that the concept of naturalness, though forward thinking when originally introduced into law and policy, does not provide adequate guidance to protected area managers regarding when, where, and how to intervene to actively manage ecosystems, especially in the context of climate change and other environmental stressors. To meet today's conservation challenges, we need more clearly defined and attainable management goals and stewardship approaches.

In this chapter, we propose a path forward for park and wilderness conservation. We first review the need to rethink park and wilderness goals, purposes, and management approaches. We offer six fundamental principles to guide stewardship in an era of rapid, profound, and unpredictable change. We then discuss how policy and science can support protected area conservation and effective stewardship.

The Trouble with Naturalness Revisited

The original idea behind the establishment of parks and wilderness was to protect certain lands from development, to maintain them in their natural state. For much of the twentieth century, the goal of preserving some places in their natural state was quite revolutionary. Naturalness requires that humans restrain their activities, that ecosystems be preserved for the sake of nonhuman species. But as the conservation challenge shifted from just preventing development to include the day-to-day stewardship of protected areas, the adequacy of naturalness as a guiding principle waned. One of the key challenges that park and wilderness managers now face is knowing where, when, and how to intervene in ecosystem processes in response to human impact and global change. Current agency policy, which directs managers to preserve or restore natural conditions, does not provide adequate guidance for meeting this challenge.

Some of the problems with naturalness reflect the different and often conflicting meanings of the term. Three primary meanings of natural are as follows:

- *Lack of human effect*. Places with little apparent human impact have sometimes been called pristine. The goal here is to preserve places such that the imprint of human activities is low or minimal.
- *Freedom from intentional human control*. Where nature is not intentionally controlled it is often considered self-willed, a concept that is often captured in the terms *wild* and *untrammeled*. Managing for self-willed nature involves human restraint. It requires

hands-off management and the absence of human manipulation of ecosystems.

- *Historical fidelity*. Historical fidelity is the goal of preserving ecosystems in states similar to those that existed in the past, with similar species composition and ecological processes (Higgs 2003). The goal here is to retain the basic ecosystem features valued when the area was designated as a protected area.

For much of the twentieth century, it was assumed that these three meanings were congruent, that ecosystems could be preserved in a pristine state without intentionally manipulating them (at least not much) and that maintaining the pristine would also preserve historical fidelity. But this notion has been undermined by advances in ecological knowledge regarding the dynamism of ecosystems and the prevalence of human impact and directional change. Ubiquitous human impact means that, in some cases, human impact can be minimized only through active manipulation of ecosystems, or intentional human control. Thus, two of the meanings of *naturalness* are in direct opposition (Cole 2000). Moreover, even in the absence of active management, disturbance occurs, succession proceeds, and climates and ecosystems change. Historical fidelity and lack of human manipulation are also then increasingly divergent.

Knowledge about the myriad ways in which aboriginal peoples have shaped park and wilderness ecosystems further challenges idealized notions of parks and wilderness as areas devoid of human impact. Indigenous peoples' activities have influenced most ecosystems for thousands of years. In the context of today's global environmental changes, all places will be substantially affected by human activities for the foreseeable future. At the same time, limiting modern human impact remains critically important for protected area conservation.

Much of this book has been about the vagueness and lack of clarity inherent to the naturalness concept, given its divergent and ultimately conflicting meanings. Management responses as divergent as doing nothing and cutting down, piling, and burning piñon–juniper have been justified through the concept of naturalness. Without more specific management goals, management actions will continue to be discretionary, uncoordinated, and ultimately ineffective in conserving large ecosystems.

Despite its limitations, we are not suggesting that naturalness be abandoned. Naturalness will continue to provide an important touchstone for protected area conservation. The concept of naturalness is powerfully evocative, reminding Americans of their original and ongoing commitment to the conservation of nature in these special places. But global environmental change requires that we identify goals that are more useful guides for stewardship.

Although there are many worthwhile goals for parks and wilderness, in this book the following goals were explored:

- Respecting nature's autonomy (by not intervening in ecosystem processes)
- Intervening to emphasize ecological integrity
- Intervening to emphasize historical fidelity
- Intervening to emphasize resilience

When managers emphasize one or more of these goals, they will preserve much of what Americans value in protected areas: biodiversity, scenic beauty, ecosystem services, wildness, iconic or otherwise important species, and recreational opportunities. But because these goals are not always complementary, important trade-offs and choices must be made. Not all protected area values can necessarily be realized in a single location. However, these goals have the potential to succeed where naturalness fails because they do not inspire conflicting interpretations and they are specific enough to be operationalized. Because they can be clearly defined, they provide important guidance regarding when intervention is appropriate and, where it is, what actions should be taken. Outcomes of interventions can be specified that are measurable, attainable, and desirable.

Forging Ahead with Protected Area Management

Given the problems with traditional park and wilderness goals, the rapid pace of global environmental change, and unprecedented uncertainty about the future, management needs to adapt and change. In this section we present six principles that encapsulate some of the most important changes that are needed.

Provide Clarity in Purpose, Approach, and Desired Outcome

Park and wilderness managers need to articulate more clearly what they are trying to accomplish and why. When interventions are being contemplated, more specific statements of purpose are needed; the restoration of natural conditions is simply too vague. Is the intent of the intervention to return the ecosystem to some past state, or is it to preserve particular species or assemblages? Because no approach can preserve all park and wilderness values, choices must be made. Choices and trade-offs should be clearly stated and justified. Or, if intervention is to be avoided, plans and proposals need to state clearly what will be gained and lost by choosing no action.

Clear statements of purpose will help managers identify appropriate management responses, including decisions about whether to intervene and, if they do intervene, whether to emphasize restoration of historical fidelity, maintenance of ecological integrity, or managing for resilience or some other ecosystem attribute. Beyond this, clear statements of purpose are critical to the specification of desired outcomes of interventions (or lack of intervention). Clear operational objectives are needed for accountability and as a means to measure success. Given future uncertainty, it is important to learn from both successes and failures. Learning is greatly facilitated by clear statements of intended outcomes of specific management actions.

Promote Diversity and Redundancy

Given the certainty that the future is uncertain and the possibility that the next century or two is a bottleneck—a period of rapid change leading to a somewhat more stable climate—preservation of future options is of paramount importance. A conservative approach to preserving options is to adopt diverse goals and management strategies. Diversity involves intervening in some situations but not in others and varying management goals and the strategies used to reach a given goal. The goals of wildness, historical fidelity, ecological integrity, and resilience overlap in many instances, and multiple goals might be appropriate in some places. Redundancy is also important. If and when interventions are undertaken, they should occur across a range of environmental situations, with multiple replications. This will spread risk and promote the opportunity to learn what does and does not work. The current diversity of goals and strategies often reflects personal preference, available resources, political pressure, and lack of coordination rather than deliberate efforts to plan for interventions in ways that consider multiple goals and values, multiple scales, boundary effects, and landscape and regional conservation efforts. Planned, purposeful diversity is necessary to systematic learning and adaptation to change. Planned diversity requires some centralized direction but could largely be coordinated through horizontal networks of protected areas and adjacent lands.

Plan at Multiple Scales

Management interventions require site-specific planning. But, as noted earlier, the spatial scale of planning also must be enlarged and diversity and redundancy must be coordinated at multiple scales. Large-scale planning is important to attempts to maximize future options by having different goals and management strategies in different places (Cole 2000) and by planning for a system of protected areas at regional scales. Planning and management must happen at multiple and appropriate scales, at scales that often depart from convention (Kareiva et al. 2008). In addition, the small size and increasing isolation of most parks and wilderness demand more attention to the matrix surrounding each area. Institutional boundaries must become much more permeable (Knight and Landres 1998). Managers must figure out how to effectively work across jurisdictional boundaries. Cooperation across boundaries means much more than information sharing; it requires collaborative goal setting and adaptation. Moreover, private landowners need to be brought into this dialogue in a way that preserves livelihood options for rural residents while also integrating critical habitat into conservation plans. Protected areas nested in adaptive landscapes will be especially important as species migrate to new ranges in response to shifts in temperature and precipitation.

Encourage Flexibility and Adaptability

Planning approaches must reflect increased uncertainty about the future. Planning must occur at multiple temporal scales and include long-term and short-term goals and allow for flexibility in adjusting goals and management actions. What appear to be realistic future options may later prove unrealistic. At the same time, new options will continually appear. Management approaches should adapt to what is learned from deliberate experimentation and effectiveness monitoring. To respond to changing public values, scientific knowledge, ecological change, and emerging management options, planning must become more flexible and adaptive.

The concept of desired future conditions—central to current park and wilderness planning—implies an understanding of alternative future states to choose from, the costs and benefits of different alternatives, the resources needed to achieve each state, and the likelihood of success. As change and uncertainty increase, these conditions are less likely to be met. Attempts to achieve long-term objectives, as conditions change, can lead to loss of resilience and long-term ecological degradation. The alternative is adaptive management and reliance on continuous learning and rigorous monitoring, regularly revisiting objectives and management decisions, and changing them as knowledge advances and uncertainty retreats (Folke et al. 2002).

Within an adaptive management framework and in the context of rapid change, prioritization and triage become particularly important. As with healthcare decision making, triage suggests that the level and urgency of intervention can be assessed depending on the degree of threat, likelihood of successful intervention, and relative value of the conservation asset under consideration (Hobbs et al. 2003). As with healthcare, individual decisions are made within a broad prioritization framework but can be revised on the basis of new evidence, changing understanding, or new technological advances.

Consider When to Look to the Future Instead of the Past

Traditionally, management interventions have been limited to mitigating threats and restoring native species and ecological processes, using past conditions as a guide. But as basic environmental conditions shift under climate change, managers need to think about interventions that look forward as well as backward in time (see Keane et al. 2008). Although restoration will undoubtedly be appropriate in many places, if conditions change too much, resisting a future that is quite different from the past may be futile and counterproductive. Increasingly, managers will need to consider actions that assist ecosystems in adapting to new conditions and, in some cases, transforming to new types of systems to avoid catastrophic degradation.

Balance Bold Action with Humility and Restraint

The threat of global environmental change to park and wilderness ecosystems creates a sense of urgency and a need for bold action. Indeed, the primary thesis of this book is that there is a need for new and innovative approaches to park and wilderness stewardship. However, the problems of uncertainty and unintended consequences also argue for humility and restraint and the need for careful analysis and inclusive decision making before action. The need for immediate action may or may not be urgent. What is urgent is dialogue between scientists, managers, and the public about park values and purposes and about appropriate approaches for realizing objectives. Thus, transparent processes for deliberating about goals, strategies, and trade-offs must be in place before management intervention.

Making Policy More Effective

Policy can either constrain or enable the institutional changes described here. Because policy dramatically influences what types of interventions are (or are not) pursued, careful consideration of existing and future policy options is critical.

The laws that guide protected area management, though visionary, are also in many ways outdated. Environmental policy in the United States is founded on the equilibrium paradigm, the assumption that ecosystems are relatively stable and static and that scientific certainty is possible (Thrower 2006). Karkkainen (2002: 197) argues that decades of environmental laws have not taken into account the inherent uncertainty and unpredictability of ecological change, and therefore current management and policy are "poorly matched to the challenge of managing ecosystems as complex systems." Current policies were not developed in the context of climate change or other environmental stressors, nor did they account for the need for flexible administrative goals. At the statutory level, key goals, such as biodiversity conservation and resilience, are largely absent for most protected areas in the United States (with the notable exceptions of the 1973 Endangered Species Act and the 1997 National Wildlife Refuge System Improvement Act).

Statutory and regulatory guidance for parks and wilderness is often vague and ambiguous, providing little specific direction for protected area management. Naturalness is inscribed in most protected area policies but is rarely, if ever, defined with any specificity. The ambiguous meaning of *naturalness* provides protected area managers with considerable discretion. If *naturalness* has multiple meanings, none of which are explicit or prioritized, many different management actions can be justified in the context of current policy. Such broad discretion sometimes means that goals for protected areas shift with personnel changes and political winds and differ across landscapes and protected areas. Although such discretion provides for flexibility and site-specific decision making, it also results in incompatible goals on adjacent parcels and lack of coordination at the landscape and regional scale, where many important ecological processes occur.

In addition to ambiguity in existing policies, multiple mandates have been articulated in law and regulation but rarely prioritized. This policy layering, whereby mandates from the Endangered Species Act and the Clean Water Act are added to the original goals of the National Park Service's Organic Act, for example, sends managers in an ever-expanding number of directions. With the exception of the national wildlife refuges, for which goals are prioritized at the statutory level, very little attention has been devoted to prioritizing goals at the system, agency, or unit level.

Articulate More Specific Goals within Agency Policy

Managing agencies should increase the specificity of the policies they articulate to implement legislative mandates. Most legislative mandates are sufficiently broad to allow implementing regulations to evolve to meet changing needs (Thrower 2006). According to Kareiva et al. (2008: 9–28), "managers can use existing legislative tools in opportunistic ways" to address environmental change. Within the current confines of protected area legislation, managing agencies could develop more specific policies based on the divergent meanings of *naturalness* and articulate diverse yet distinct stewardship goals based on the concepts described in this book (autonomous nature, historical fidelity, resilience, and ecological integrity) or others. For example, the Forest Service's 2008 Interim Directive on Ecological Restoration and Resilience directs managers to "maintain the adaptive capacity of ecosystems" and develop goals and objectives that consider "current and likely future ecological capabilities" and "future changes in environmental conditions." Diverse goals within or between protected areas would need to be coordinated to account for boundary effects and necessary redundancies and thus would need to be nested within efforts to plan at landscape and regional scales. Policies could prioritize specific goals for individual protected areas or sites within protected areas, because multiple goals may be incompatible.

Balance Flexibility, Specificity, and Accountability in Policy and Planning

Managers and other decision makers need to carefully negotiate tensions between specific goals, flexibility to adapt, and accountability to the American public. Increasing the specificity of goals will necessarily limit management discretion. But in many ways adaptive, responsive management depends on some degree of management discretion. Managers must have the latitude to respond quickly in the face of unanticipated change. In addition to this tension, adaptive management as an "ongoing experiment poses a profound challenge to our legal system because it undermines a core principle of procedural and substantive fairness: finality" (Tarlock 1994: 1140). With regard to National Forests, Burchfield and Nie (2008) argue for the need to balance adaptable planning and enforceable standards. One way to address this tension is to narrow the sideboards through more specific goals while providing substantial latitude within those sideboards. There is a critical need to develop policies and decision-making mechanisms that include clearly specified and prioritized goals and mechanisms for quality control and public accountability, while also enabling nimble, flexible experimentation. Scenario planning, as described in Chapter 13, with significant public involvement is one way to combine accountability with adaptability.

Address Existing Policy Constraints

In some cases, current environmental policy restricts managers' ability to pursue approaches discussed in this book. For example, in Chapter 10 it is suggested that triage may be necessary given limited resources and rapid change. Triage suggests focusing management efforts on some imperiled species, ecosystems, and processes over others. However, the Endangered Species Act requires that all threatened and endangered species be recovered, and there are no provisions that assist agencies in prioritizing one species' survival over another. Agency policies regarding invasive species may not account for possible range changes caused by shifting climatic conditions (Scott and Lemieux 2005). Such policies might define a migrating species as invasive and recommend that species be eradicated. Chapter 5 discusses how the National Environmental Policy Act requires that agencies commit to a specific course of action for a particular period of time. Policies that require lasting decisions may limit flexibility and adaptation in the face of unexpected change. As managers and the public consider new goals and approaches, existing statutes and regulations may need to be amended to allow for innovative and adaptive management. At the same time, some statutes provide constraints that might be seen as desirable. For example, the mandate that wilderness areas remain untrammeled may restrict management intervention in such places. For advocates of autonomous nature, the Wilderness Act then provides a desirable constraint. For those who want to see some intervention in wilderness, the Wilderness Act might need to be amended.

Opening up political dialogue about cherished environmental policies involves some level of risk, because the changes ultimately made might be worse than the current situation and might favor development over conservation. However, the example of the National Wildlife Refuge Improvement Act (1997) demonstrates that at times Congress has been willing to provide more specific statutory guidance and also increase the level of protection afforded to protected areas. The new refuge statute has been touted as the "most expansive ecological mandate in U.S. public land law" because it incorporates biodiversity conservation to a far greater extent than any other statute guiding protected area management in the United States (Fischman and Meretsky 2004: 940).

Prioritize Protected Area Goals

Policymakers and agency decision makers should also consider how to prioritize goals for protected areas, so that managers are not constantly faced with difficult trade-offs and little guidance. Goals could be prioritized through statutory mandates, within agency policy, or at the regional or unit level. Prioritization of goals would provide managers with much clearer guidance regarding what types of interventions they should pursue. As part of this process, new goals and values should be considered. The National Wildlife Refuge System Improvement Act (1997) provides a possible model for incorporating biodiversity conservation as a critical stewardship goal. The Refuge Improvement Act states that refuges should "ensure that the biological diversity, integrity, and environmental health of the system are maintained for the benefit of present and future generations of Americans." The act provides a hierarchy of priorities to assist managers faced with competing goals. Goals could also be prioritized through coordinated efforts across agencies and individual units in such a way that provides for both specific direction and flexibility to adapt to change.

Commission a Review of Current Protected Area Policy

To be effective, policy should draw on the wisdom of diverse stakeholders and experts. In the spirit of the Leopold report, policymakers might commission a panel, such as a National Academy of Sciences Review, to consider the appropriateness of current laws and policies affecting management of parks and wilderness and to make recommendations for change. Such a panel should include agency staff, scientists, policymakers, and stakeholders. The panel could consider a number of possible policy changes, such as how to articulate more specific goals for parks and wilderness, how to develop planning regulations that facilitate more adaptive and flexible planning while ensuring accountability, and whether or not to pursue statutory requirements for working with adjacent landowners, as was done in the National Wildlife Refuge System Improvement Act.

Embrace Public Debate

Public debate is critical to effective development of plans, goals, and priorities for protected area conservation. Public understanding of both ecological change and management interventions may initially be low (Bosworth et al. 2008). Members of the public may be attached to specific landscapes and ecosystems and resist the transformation of such places. The consequences of climate change, especially where there is visible, catastrophic loss (as with widespread forest die-offs), may result in political backlash against managing agencies. Many innovative proposals, such as assisted migration (which may seem overly manipulative) or triage (especially if beloved species or ecosystems are not prioritized), will be controversial. To build the political capital and public support critical to adaptive stewardship, managers need to embrace public debate and channel public interest into engaging public involvement processes.

Develop Dynamic, Participatory Public Involvement Processes

Public involvement in protected area decision making must be dynamic, participatory, and inclusive. Policy decisions are fundamentally political decisions; they are "scientifically informed value judgments" (Tarlock 1994: 1133). Science provides information about possible options, consequences, trade-offs, and resources, but a broader citizenry will ultimately determine the future of parks and wilderness. Public involvement must be inclusive, engaging diverse stakeholders in a meaningful dialogue. Public debate about the future of protected areas will be contentious. But conflict calls for participatory processes such as collaboration, transactive planning, and multiparty monitoring, processes that provide meaningful opportunities for interested parties to become involved in policy and management. A broad range of values, including recreational values, ecosystem services, and economic interests, should be considered alongside the values highlighted in this book, such as biodiversity and resilience. Participatory planning at the regional and unit scales is necessary to work through trade-offs and difficult choices. Public involvement in scenario planning can help build support if goals must be modified (Baron et al. 2008). If trade-offs can be deliberated in anticipation of change, both managers and the public will feel more comfortable responding decisively in the face of unexpected change. In the context of adaptive management, planning is ongoing, and planners will need to develop tools to keep the public engaged in long-term dialogue. Compared with the past, when most public involvement meant allowing citizens to comment on proposed actions and plans, the public involvement of the future must become more dynamic, with a pace and process to match an increased need for flexible and adaptive decision making. Managers and scientists also need to develop effective ways of describing both scientific uncertainty and the need for active management interventions.

Learning from Science

Scientific research is an essential piece of the puzzle, and such research should be part of the broader dialogue about institutional and policy change. In many places throughout this book, we have argued that there is much uncertainty about the future and how environmental change will affect protected areas and the ability to manage for particular goals. Uncertainty suggests an ongoing need to improve scientific understandings of the ecosystems and species within and beyond protected areas so there is a stronger foundation of knowledge on which to base management decisions (Sutherland et al. 2004; Sutherland 2006). There remains a pressing need for continued and increased collection of basic information on species distributions and dynamics, ecosystem structure and function, and how environmental changes act and interact to modify them. Such research would be needed even in the absence of rapid environmental change, but it is more critical now. Another key challenge is to set up and maintain effective and efficient monitoring programs that focus on key variables and provide direct triggers for management intervention. But given that funds for research are perennially limited, it is important to prioritize research needs, as has been done by Soulé and Orians (2001) and Wu and Hobbs (2002). This section describes some of the most pressing areas of research.

Understanding How to Move Forward Given Uncertainty

Traditionally, research has attempted to reduce the uncertainty involved in decision making. Better understanding of how systems work should theoretically lead to more accurate models of those systems and hence an improved prediction capability. However, it is becoming increasingly clear that although there are situations in which this is possible, the complexity and nonlinearity of ecosystems and their dynamics often make prediction virtually impossible. Funtowicz and Ravetz (1990) identify realms of known, unknown, and unknowable, as defined by the degree of uncertainty about the likelihood and outcomes of particular events, and suggest that very different approaches are needed in each case. Traditional scientific methods, using reductive and frequentist approaches, are useful only where uncertainty is fairly small. As uncertainty increases, more subjective approaches and techniques such as scenario analysis become more appropriate. Ecology has been slow to embrace alternative and integrative approaches to complex environmental issues. There is a pressing need for building and testing such approaches in the context of protected area management and developing decision tools that allow managers to incorporate uncertainty into decision making.

Anticipating Thresholds

Part of the uncertainty discussed here relates to increasing recognition of the importance of thresholds in ecological systems. Threshold dynamics mean that there are nonlinearities in the way systems respond to changes in environmental variables, such that a small change in the driving variable can push the system across a threshold and lead to dramatic changes in ecosystem properties. Threshold phenomena are well documented in freshwater systems and are increasingly discussed in terrestrial ecosystems (Hobbs and Suding 2008). However, the ability to recognize and predict thresholds before they are crossed has eluded ecologists. Thresholds are generally observed only after they have been crossed. With the increasing pace of environmental change, the likelihood of threshold dynamics increases, and an improved understanding of how to recognize and anticipate such dynamics is urgently needed (Suding and Hobbs 2009).

Understanding Resilience

In various chapters throughout this book, the concept of resilience has been discussed as a valid and useful goal for conservation management, especially in the context of rapid environmental change. Although there has been much progress in moving from metaphor to measurable characteristics of resilience (Carpenter et al. 2001), more work is needed to fully operationalize resilience. More clarification is needed about exactly what it is and how managers can recognize and achieve it (Chapin et al. 2003; Brand and Jax 2007).

Using Historic Information to Manage for an Uncertain Future

The past has loomed large in current conservation and restoration approaches, which often aim to preserve or restore historic conditions of ecosystems. However, it is increasingly apparent that our understanding of the past is constantly evolving, and therefore interpretation of the past for future management also needs to change. This is particularly the case for scientific understandings of the role of indigenous peoples in ecosystem management and modification. Furthermore, it is increasingly apparent that the environments and ecosystems of the future will not necessarily have any past analogs for managers to refer to for guidance (Hobbs et al. 2006; Williams and Jackson 2007). How then do protected areas retain some anchor to the past while dealing with the future? In particular, how do place-based conservation policy and management need to adapt to changing environments and species distributions? This question clearly transcends basic ecology and involves deep philosophical and political issues and therefore demands an interdisciplinary approach to finding ways through the intellectual quagmire that might result.

Building Adaptive Institutions and Effective Planning Processes

Because uncertainty about how to develop effective future institutions, policies, and practice is as pronounced as uncertainty about how to protect future ecosystems, social science must augment the biological and physical sciences. An improved understanding of public views of protected area goals and values, current and future ecological change, and possible management intervention will assist managers in decision making. Social science can contribute an improved understanding of people's ability to make decisions in the context of uncertainty and help develop tools for making trade-offs (Kareiva et al. 2008). Just as ecological systems need to be monitored, we need to monitor institutions, policies, and practices to ensure they are effective and contribute to resilience. We need research on how to enhance the adaptive capacity of managing institutions (Joyce et al. 2008), how to plan and manage across boundaries, how to balance adaptability and accountability, and how to involve the public in dynamic, ongoing decision making.

Rising to the Challenge

The ecological upheavals wrought by climate change, invasive species, pollution, and habitat fragmentation greatly threaten the values for which parks and wilderness were established. This book argues that current agency policies based on maintaining naturalness in protected areas are inadequate given the challenges of responding to these threats. The concept of naturalness does not provide sufficient guidance; it has multiple meanings that are often in conflict. Consequently, the path forward, especially in the context of global environmental change, lies in the development of new goals that can be operationalized and new approaches that are effective in a world of change and uncertainty.

In this chapter, we have described principles for managing protected areas and suggested ways to strengthen policy and science. We believe it is time for managers and the American public to move toward more specific stewardship goals. Effectively responding to ecological change and uncertainty will require fundamental shifts in institutions, policies, and practices, shifts that promote innovation, experimentation, and a level of flexibility and agility uncommon in large government bureaucracies. These changes are significant but necessary. We have every expectation that protected area managers, scientists, policymakers, and the American public can rise to the challenge and move forward on protected area conservation. The challenge of global environmental change represents an opportunity for renewed investment in parks and wilderness, a careful rethinking of the purposes of such areas, and a revitalized way of conserving biodiversity in these special places.

REFERENCES

Baron, J. S., C. D. Allen, E. Fleishman, L. Gunderson, D. McKenzie, L. Meyerson, J. Oropeza, and N. Stephenson. 2008. National parks. Pp. 4-1–4-68 in S. H. Julius and J. M. West, eds. *Preliminary review of adaptation options for climate-sensitive ecosystems and resources: A report by the U.S. Climate Change Science Program and the Subcommittee on Global Change Research.* U.S. Environmental Protection Agency, Washington, DC.

Bosworth, D., R. Birdsey, L. Joyce, and C. Millar. 2008. Climate change and the nation's forests: Challenges and opportunities. *Journal of Forestry* 106:214–221.

Brand, F. S., and K. Jax. 2007. Focusing meaning(s) of resilience: Resilience as a descriptive concept and a boundary object. *Ecology and Society* 12(1):23. Retrieved September 21, 2009 from www.ecologyandsociety.org/vol12/iss1/art23/.

Burchfield, J., and M. Nie. 2008. *National forest policy assessment.* University of Montana, College of Forestry and Conservation, Missoula.

Carpenter, S., B. Walker, J. M. Anderies, and N. Abel. 2001. From metaphor to measurement: Resilience of what to what? *Ecosystems* 4:765–781.

Chapin, F. S. III, T. S. Rupp, A. M. Starfield, L. DeWilde, E. S. Zavaleta, N. Fresco, J. Henkelman, and A. D. McGuire. 2003. Planning for resilience: Modeling change in human–fire interactions in the Alaskan boreal forest. *Frontiers in Ecology and the Environment* 1:255–261.

Cole, D. N. 2000. Paradox of the primeval: Ecological restoration in wilderness. *Ecological Restoration* 18(2):77–86.

Fischman, R. L., and V. J. Meretsky. 2004. Managing biological integrity, diversity, and environmental health in national wildlife refuges: An introduction to the symposium. *Natural Resources Journal* 44:931–988.

Folke, C., S. Carpenter, T. Elmqvist, L. Gunderson, C. S. Holling, B. Walker, J. Bengtsson, et al. 2002. *Resilience and sustainable development: Building adaptive management in a world of transformation.* Series on Science for Sustainable Development No. 3. The Environmental Advisory Council to the Swedish Government, Stockholm.

Funtowicz, S. O., and J. R. Ravetz. 1990. *Global environmental issues and post-normal science.* London Council for Science and Society, London.

Higgs, E. 2003. *Nature by design: People, natural process, and ecological restoration.* The MIT Press, Cambridge, MA.

Hobbs, R. J., S. Arico, J. Aronson, J. S. Baron, P. Bridgewater, V. A. Cramer, P. R. Epstein, et al. 2006. Novel ecosystems: Theoretical and management aspects of the new ecological world order. *Global Ecology and Biogeography* 15:1–7.

Hobbs, R. J., V. A. Cramer, and L. J. Kristjanson. 2003. What happens if we can't fix it? Triage, palliative care and setting priorities in salinising landscapes. *Australian Journal of Botany* 51:647–653.

Hobbs, R. J., and K. N. Suding, eds. 2008. *New models for ecosystem dynamics and restoration*. Island Press, Washington, DC.

Joyce, L. A., G. M. Blate, J. S. Littell, S. G. McNulty, C. I. Millar, S. C. Moser, R. P. Neilson, K. O'Halloran, and D. L. Peterson. 2008. National forests. Pp. 3-1-3-127 in S. H. Julius and J. M. West, eds. *Preliminary review of adaptation options for climate-sensitive ecosystems and resources: A report by the U.S. Climate Change Science Program and the Subcommittee on Global Change Research*. U.S. Environmental Protection Agency, Washington, DC.

Kareiva, P., C. Enquist, A. Johnson, S. H. Julius, J. Lawler, B. Petersen, L. Pitelka, R. Shaw, and J. M. West. 2008. Synthesis and conclusions. Pp. 9-1-9-66 in S. H. Julius and J. M. West, eds. *Preliminary review of adaptation options for climate-sensitive ecosystems and resources: A report by the U.S. Climate Change Science Program and the Subcommittee on Global Change Research*. U.S. Environmental Protection Agency, Washington, DC.

Karkkainen, B. 2002. Collaborative ecosystem governance: Scale, complexity, and dynamism. *Virginia Environmental Law Journal* 21:189–243.

Keane, R. E., L. M. Holsinger, R. A. Parsons, and K. Gray. 2008. Climate change effects on historical range and variability of two large landscapes in western Montana, USA. *Forest Ecology and Management* 254:375–389.

Knight, R. L., and P. B. Landres. 1998. *Stewardship across boundaries*. Island Press, Washington, DC.

Scott, D., and C. Lemieux. 2005. Climate change and protected area policy and planning in Canada. *The Forestry Chronicle* 83:696–703.

Soulé, M. E., and G. H. Orians, eds. 2001. *Conservation biology: Research priorities for the next decade*. Island Press, Washington, DC.

Suding, K. N., and R. J. Hobbs. 2009. Threshold models in conservation and restoration: A developing framework. *Trends in Ecology and Evolution* 24:271–279.

Sutherland, W. J. 2006. Predicting the ecological consequences of environmental change: A review of the methods. *Journal of Applied Ecology* 43:599–616.

Sutherland, W. J., A. S. Pullin, P. M. Dolman, and T. M. Knight. 2004. The need for evidence-based conservation. *Trends in Ecology and Evolution* 19:305–308.

Tarlock, D. 1994. The nonequilibrium paradigm of ecology and the partial unraveling of environmental law. *Loyola of Los Angeles Law Review* 27:1121–1144.

Thrower, J. 2006. Adaptive management and NEPA: How a nonequilibrium view of ecosystems mandates flexible regulation. *Ecological Law Quarterly* 33:871–895.

Williams, J. W., and S. T. Jackson. 2007. Novel climates, no-analog communities, and ecological surprises. *Frontiers in Ecology and the Environment* 5:475–482.

Wu, J., and R. Hobbs. 2002. Key issues and research priorities in landscape ecology: An idiosyncratic synthesis. *Landscape Ecology* 17:355–365.

GREGORY H. APLET is senior forest scientist in the Wilderness Society's Ecology and Economics Research Department in Denver, Colorado. His current work focuses on the ecology and management of wildland fire and the management of landscapes to sustain wildland values and services in the face of global environmental change.

F. STUART CHAPIN III is an ecosystem ecologist whose research addresses the sustainability of ecosystems and human communities on a rapidly changing planet. This work emphasizes the impacts of climate change on Alaskan ecology, subsistence resources, and indigenous communities as a basis for developing climate change adaptation plans.

DAVID N. COLE is a geographer at the USDA Forest Service's Aldo Leopold Wilderness Research Institute in Missoula, Montana. He has conducted a wide variety of research relevant to the stewardship of wilderness areas, particularly in recreation and restoration ecology, wilderness visitor research, development of planning frameworks, and defining wilderness goals and purposes.

PAUL DEPREY was chief of resource management at Joshua Tree National Park. His National Park Service career has focused on encouraging parks to understand and appreciate large-scale changes. He has recently been posted to Superintendent of WWII Valor of the Pacific National Monument, based in Pearl Harbor, Hawaii.

DAVID M. GRABER is chief scientist for the Pacific West Region of the National Park Service, where he plans strategic conservation research and ensures that science is used effectively in national park management. Previously he worked on species–habitat relationships and the establishment of a National Park Service inventory and monitoring program.

ERIC S. HIGGS is a professor in the School of Environmental Studies at the University of Victoria, Canada. His work focuses on conceptual foundations of restoration, historical ecology and landscape change in mountainous regions, and protected areas management. He is the author of *Nature by Design: People, Natural Process, and Ecological Restoration* (MIT Press, 2003).

RICHARD J. HOBBS is an Australian professorial fellow at the University of Western Australia. His research interests span basic ecology, restoration ecology, landscape ecology, and conservation biology, and he focuses on the management and repair of altered ecosystems. He is editor in chief of the journal *Restoration Ecology*.

PETER LANDRES is an ecologist at the Aldo Leopold Wilderness Research Institute, part of the USDA Forest Service Rocky Mountain Research Station. His recent research includes helping to develop an interagency strategy to define and monitor trends in wilderness character across the U.S. National Wilderness Preservation System.

CONSTANCE I. MILLAR is a research paleoecologist with the USDA Forest Service, Sierra Nevada Research Center, Berkeley, California. Her research focuses on responses of high-elevation forests to historic climate variability, with current interest extending to rock glaciers, American pika, and alpine vegetation monitoring.

DAVID J. PARSONS is director of the USDA Forest Service's Aldo Leopold Wilderness Research Institute in Missoula, Montana. He is a plant ecologist interested in science to support policy and management of protected areas. He focuses on vegetation, fire and other disturbances, and the impacts and management of recreation use.

JOHN M. RANDALL is the associate science director for South Coast and Deserts with The Nature Conservancy. From 2004 to 2009, he led the conservancy's Global Invasive Species Team. His work focuses on invasive species threats to biological diversity, issues surrounding the development of renewable energy infrastructure, and conservation planning.

NATHAN L. STEPHENSON is a research ecologist with the U.S. Geological Survey's Western Ecological Research Center, stationed in Sequoia and Kings Canyon national parks. He is particularly interested in effects of rapid climatic changes on forests and how protected area managers can adapt to an uncertain, but certainly unprecedented, future.

KATHY A. TONNESSEN is a research coordinator with the National Park Service, Rocky Mountains Cooperative Ecosystem Studies Unit, located at the University of Montana. She works on natural and cultural resource management and science issues of concern to Rocky Mountain national parks and has expertise in the effects of air pollution on natural ecosystems.

LEIGH A. WELLING is the climate change coordinator for the National Park Service. She has a Ph.D. in oceanography from Oregon State University and has conducted paleoclimate research in the eastern equatorial and northeast Pacific. She coordinates servicewide efforts in adaptation planning and development of decision support tools for resource protection.

PETER S. WHITE is a professor of biology and director of the North Carolina Botanical Garden at the University of North Carolina. His work focuses on patterns of biodiversity, ecosystem dynamics and restoration, and conservation issues. He and his students often work in Great Smoky Mountains National Park.

STEPHEN WOODLEY is the chief ecosystem scientist for Parks Canada, where he works on a number of issues related to protected areas, including developing techniques for monitoring and assessment of ecological integrity, ecological restoration, fire, and sustainable forestry. He is a member of the World Conservation Union's Commission on Protected Areas.

LAURIE YUNG is a conservation social scientist and director of the Wilderness Institute at the University of Montana. Her research focuses on how rural communities negotiate environmental change, landownership change and landscape conservation, and the social and political aspects of climate change adaptation.

ERIKA S. ZAVALETA is a terrestrial ecologist in the Environmental Studies Department at the University of California, Santa Cruz. She studies community and ecosystem ecology, with emphasis on the drivers and implications of biodiversity change for ecological and human well-being. Recent work focuses on grassland and woodland conservation in California.

Island Press | Board of Directors